An Entirely Synthetic Fish

An Entirely Synthetic Fish

How Rainbow Trout

Beguiled America and

Overran the World

Anders Halverson

Yale University Press

New Haven and London

Published with assistance from the Louis Stern
Memorial Fund.

Designed by Sonia Shannon
Set in Palatino type by Tseng Information Systems, Inc.,
Durham, North Carolina.
Printed in the United States of America.

The Library of Congress has cataloged the hardcover
edition as follows:
Halverson, Anders, 1969–
An entirely synthetic fish : how rainbow trout beguiled
America and overran the world / Anders Halverson.
          p.   cm.
 Includes bibliographical references and index.
 ISBN 978-0-300-14087-3 (alk. paper)
1. Rainbow trout. 2. Introduced fishes—United States.
3. Rainbow trout industry—United States—History.
I. Title.
QL638.S2H187 2010
639.3′757—dc22

                              2009036200

ISBN 978-0-300-14088-0 (pbk.)

A catalogue record for this book is available from the

British Library.

For Ginna

Contents

Foreword

PATRICIA   NELSON   LIMERICK

"Complaints are everywhere heard . . . that the public good is disregarded in the conflict of rival parties," founding father James Madison wrote in a passage that sounds as if he planned a second career as an environmental policy analyst. When Madison expressed this thought in *The Federalist Papers*, however, he did not have recreational fishermen, advocates of biodiversity, state and federal resource managers, or business owners in resort communities to cite as his examples. But if we could summon Madison for a twenty-first-century site visit to any popular trout-fishing stream, he would arrive with a well-designed conceptual framework ready to apply. "As long as the reason of man continues fallible, and he is at liberty to exercise it," Madison wrote, "different opinions will be formed. As long as the connection subsists between his reason and his self-love, his opinions and his passions will have a reciprocal influence on each other." And with reason and self-interest thus entangled, Madison concluded, "the latent causes of faction are thus sown in the nature of man."

The qualities of human nature that worried Madison have

come to thrive in environmental disputes today. And yet, facing off against the generous and open spirit of Anders Halverson's *An Entirely Synthetic Fish*, faction and its comrade-in-arms, un-reasoned passion, back down and come close to apologizing for all the trouble they have caused. Tracing the massive enterprise that produced rainbow trout in hatcheries and dispersed them worldwide, Halverson explores a case study full of implica-tion for hundreds of other environmental matters. In truth, his approach sets a model for taking on topics of contention and conflict. This book combines a respectful and empathetic explo-ration of the beliefs and attitudes that guided this extraordinary project in the rearrangement of nature with a well-informed and clear presentation of scientific findings on the project's conse-quences.

With that combination, this book rewards readers in two equally significant ways. First, it entertains us with stories of in-trinsic interest and even mind-stretching improbability. Second, it invites us to be smarter and more congenial citizens, more inclined to think productively about our environmental chal-lenges and dilemmas, and more prepared to rise above faction and return to regarding "the public good."

The stories are indeed extraordinary. Post–Civil War trans-porters of fish eggs endured exhausting railroad trips, denied sleep for a week as they added ice and water to keep the eggs alive. After World War II, the California Department of Fish and Game took the practice of fish transportation in an even more improbable direction, populating the once fish-free lakes of the Sierra Nevada by acquiring surplus military planes and launch-ing cascades of fish in free fall into the water. Then, after years of purposeful stocking of hatchery fish, alarm over the threat posed to endangered native fish led, in many locations, to an about-face, with fish eradication projects diametrically reversing the meanings of success and progress. And, in the most intriguing plot twist of all, the fish have taken over the action, as native fish and introduced fish get down to business and produce a new kind of fish: hybrids whose muddled DNA challenges laws and policies designed to deal with a considerably more predictable and tractable material world. Never a simple matter, the goal of preserving an endangered species becomes dramatically more complicated when the species itself does not hold still and retain a steady identity.

And, at this point, Halverson's stories begin their own move toward hybridity, emerging as factually accurate recountings that also present the DNA markers of parables and fables. While readers who simply want to savor and enjoy the ride will have a fine time, the stories are ready to deliver a much greater and more consequential reward. People can treat this book as exercise equipment for the intellect and imagination, a device to strengthen and stretch the mind's powers to reflect and to analyze in the most useful way.

Here's an example of the usefulness I have in mind.

In the early 1990s, I spoke at the annual conference of the Western Association of Fish and Wildlife Agencies and met a memorable group of professionals. In truth, with his very biodiverse ark, Noah could be cast as the father and founder of this particular profession. The audience was made up of state officials charged with managing a vast range of diverse creatures: furred and scaly, colorful and dull, charismatic and unnerving, endangered and abundant, game and non-game. In the late afternoons, a good-hearted person, who himself seemed to be a hybrid of sociologist and therapist, conducted sessions to help these managers reduce the stress and tension of dealing with the most vexing and trying organisms of all: the human constituents (a.k.a. factions) whose dreams, desires, and demands made the professional lives of the managers so challenging.

At the beginning of one session, our sociologist-therapist-hybrid leader instructed us to draw pictures of vicious circles that characterized our professional lives. Professors are not bereft of material for such an assignment, and I sketched my own poignant diagram of one recent loop of escalating university miscommunication. But then, as we settled in to contemplate one another's diagrams, my companions' troubles trumped and dwarfed my own. Their vicious circles all followed a path of initial constituent demands, to which the managers responded as helpfully as they could. From there, the circles diverged between successful responses to demands that then led the constituents to believe that they could get what they wanted (usually a robust population of game fish, deer, or wildfowl) and thus led them to demand even more, *or*, unsuccessful responses that threw the dissatisfied constituents into a frenzy of escalating demands. Either way, with each turn of the circle, the professional lives of the wildlife managers ratcheted up another notch or two in tension and vexation. And, as the managers tried to respond to the demands of one particular faction, the demands of many other factions came at them at an unrelenting pace, as, for instance, environmental groups questioned the whole project of letting the preferences of hunting and fishing advocates set the priorities of wildlife management.

An hour of immersion in the dilemmas of fish and wildlife managers left me with a lasting recognition of the miniature scale of my own professional burdens. In this situation, as in many other encounters with the people we have charged with managing our national resources, I wished that I could add one item to my own task list: helping public servants who are so regularly and repeatedly buffeted by conflicting demands of citizens.

But how?

I would have liked to have been able to hand my new pals at the meeting of the Western Association of Fish and Wildlife Agencies, and other such convenings, a copy of this book, and to say to them, "Read this yourselves, but even more important, get it into the hands of the citizens whose dreams, desires, and demands so often land on you and the resources you manage. The more citizens who read this, the better off we all will be."

Why?

Because this book responds to a number of this nation's greatest needs.

First, in an impatient nation with a short attention span, a reckoning with the origins of our dilemmas is not everyone's idea of a valuable investment of time. Yet societies with such a reluctance to pay attention to the past display many of the same symptoms seen in unlucky individuals afflicted with amnesia. In the absence of any sense of the circumstances that brought a particular dilemma into being, the chances of dealing productively with it become very slight. Thus, a writer must be capable of unusual persuasiveness to invite people to make such a reckoning, and in this book, Anders Halverson displays that persuasiveness in spades.

Second, the capacity to present complicated scientific understandings with clarity, energy, and charm is urgently needed, and Halverson has exemplary skills in this terrain. Even the best-intentioned experts, trying to describe the ways in which transplanted rainbow trout can threaten the survival of native fish, will teeter on the edge of jargon. Halverson does not go off the edge of that cliff. On the contrary, he likens rainbow trout to "a school bus full of teenage boys in front of an all-you-can-eat buffet. There's not a lot of food left by the time the other customers get to the counter." With skillful expression like this, even the most science-averse readers will follow the story, free of anxiety or disorientation.

This is a book in which the author both advocates and demonstrates the virtue of humility. One of Halverson's great assets as a thinker and interpreter is his freedom from the smugness and unyielding certainty that characterize many who take up environmental subjects. In the epilogue, he suggests only

> that those who promote the conservation and restoration of native species should do so with a . . . sense of humility. . . . Reading through the letters and public pronouncements of the men who were most responsible for spreading nonnative species like rainbow trout throughout the world in the nineteenth century, I have been struck by the similarity of the rhetoric to those who promote native species restoration today.

>They, too, were sure they were doing the right thing
>for the world.

"Doing the right thing for the world," or, in Madison's wording, reducing the power of "the conflict of rival parties" to distract us from "the public good," does require courage, conviction, and determination. But it also requires flexibility, self-examination, and, perhaps most of all, humility. And therein lies the most valuable contribution of this book. After several centuries in which human beings have unleashed their ingenuity without this crucial tempering force, the time is upon us to let humility back into the game.

Preface

Through the pile of books, nominally important papers, and electronic cables that restlessly shift around my office, a stuffed rainbow trout occasionally surfaces. Someday I intend to hang it on the wall. It wouldn't be a particularly memorable fish to most people; it's only about ten inches long, and except for the light splash of pink down its side, it's not all that colorful. To me, though, it is an object of considerable interest.

For one thing, I made it, or at least I helped. Not the taxidermy—I let a professional do that—but the fish, the living fish, the fish that I caught one spring day in a reservoir on Colorado's Front Range. About a year and half before I caught it, I visited a Colorado Division of Wildlife fish hatchery when the managers were propagating their rainbow trout. They let me help. They showed me how to grab the ripe females and slide my thumb and forefinger along their belly toward the tail, squeezing a stream of beautiful salmon-colored eggs into a pan. Repeating the process with the battle-scarred males yielded a squirt of white milt on top. Swirling the pan around quickly mixed everything up, and within a few short seconds, fertilization was complete. The eggs were on their way to becoming some of the 9 million rainbow trout that Colorado stocks into its rivers and lakes every year.

Through telephone calls, emails, and occasional visits, I followed that lot of fish as they made their way from the hatchery to the ponds in which they were raised and, ultimately, onto a truck. After a long drive, I watched them cascade out of a tube on the back and into the lake. And shortly thereafter, I caught a couple, including the one that I sent to the taxidermist.

It's not just my role in its beginning and end, though, that so intrigues me with this stuffed trout. It's the unresolved conundrum that it represents. I grew up fishing for rainbows in Colorado. Like many anglers, I found something soulful and even spiritual about the activity, an escape from civilization and a means of connecting to the natural world. At some point, though, I became aware of a fundamental incongruity that ultimately became impossible for me to ignore.

Consider that today, about two out of every three fish swimming Colorado's waters are not native to the state. Many of them are rainbow trout, introduced for recreational anglers like me. And it's not just in Colorado. Rainbows are native only to the watersheds of the Pacific Rim, from Mexico to Kamchatka. Over the past century, though, billions of rainbow trout have been stocked into every state in the United States and every continent except Antarctica.

These fish have become a worldwide cultural phenomenon. They feature in postage stamps from Malawi to New Zealand, they are being considered as an "honorary indigenous species" in South Africa, and until recently they were the state fish of both Colorado and Utah.

What's more, the United States stocks more rainbow trout these days than at any other point in history, at least for which the data are available. State and federal agencies typically stock about 2 billion fish, about 40 million pounds, each year—enough to give every American household the makings of a nice meal. More than half of those fish are walleye. But that's just counting the numbers. Most walleye are stocked at a very small size. A better measure is total weight. And by that metric, the rainbows are tops. About 100 million of them leave the hatcheries every year, weighing about a quarter pound apiece—a total of 25 million pounds of rainbow trout dumped into America's freshwaters on an annual basis. And that doesn't even include the steelhead or the private hatcheries.

Of course you can't do something on that scale and not have an impact. Take amphibians. For my Ph.D., I studied these creatures. I didn't examine how fish stocking affected them, at least not directly. I was a molecular biologist, using DNA-fingerprinting techniques to examine questions of ecology, evolution, and conservation. I did spend most of my springs and summers hip-deep in ponds, though, and many hours in the library reading articles about aquatic ecology. And you can't spend a lot of time in either place without becoming keenly aware of fish, the kings of the watery realm. To frogs, for example, fish are fearsome predators and often vectors of disease; in some cases, they represent extinction itself. Stock a pond with fish, and you can usually bid the amphibians goodbye. Stock enough ponds, and you may even wipe out an entire species.

Because it can have such a powerful effect on amphibians and other organisms, fisheries management has become a battle-ground in recent years, especially when the Endangered Species Act becomes part of the equation. The salmon hatcheries of the Pacific Northwest have provided untold attorneys with a decent living for at least the past century. Would-be protectors of native species often sue to prevent fish and game departments from stocking fish. On the flip side, protests are not uncommon when the same managers try to protect native species by removing introduced fish. Even a reduction in the amount of fish being stocked can lead to fireworks. Anglers and the small businesses that depend on them tend to be a vocal lot.

Frequently, it is the merits of the science or the technicalities of the law that drive the arguments. I have become convinced, though, that the root of many of these disputes lies in deeply held, seldom-discussed beliefs about the rightful human place in the natural world. And so, for more than three years at the Center of the American West at the University of Colorado, I tried to unearth the roots of some of these conflicts. My travels took me from the wilderness of northern California to the National Archives in Washington, D.C., and to many other places in between. I talked to more than one hundred people including fisheries managers, anglers, fly-shop owners, historians, and biologists.

What I found was an engrossing collection of fact and opinion that more than once forced me to question my own prejudices and value systems. More important, though, what I found was a good story. And that is what I have tried to bring to the forefront in this book. Those seeking a book of pensive environmental essays or some sort of polemic may want to look elsewhere. So, too, those looking for a natural history or a how-to book. There are many talented writers who excel at all of those forms, and interested readers can find many of their works listed in the bibliography. My goal with this book is to tell a good tale, one that ultimately is not so much about the fish as it is about the humans. Personally, I'll never be able to look at a rainbow trout the same way again, whether it is on a plate, on the end of a hook, or even next to a pile of papers on my desk. My hope is that the reader will finish this book with at least some of the same sense of marvel. Finally, I have put many of the pictures and documents that I collected while researching this book online. If you wish to see them, discuss any of the issues raised here, or contact me about upcoming events, please visit http://andershalverson.com.

Acknowledgments

I am persuaded that many of our most valuable historical records and artifacts reside in the small historical societies and archives that dot the country. While I have had the pleasure of visiting only a few of them, my gratitude extends to the staff and volunteers of all such institutions for safekeeping these treasures and so selflessly sharing them with people like me. My particular thanks go to Lea Kemp at the Rochester Museum and Science Center, all the volunteers at the Shasta Historical Society, Su Garrison-Terry and Chris Kretz at the Dowling College Archives and Special Collections, Gil Bergen at the Connetquot River State Park Preserve, and especially Joyce Higgins and Marge Reed at the Charlestown Historical Society.

I am likewise indebted to those people who so willingly shared their knowledge, memories, and private collections of letters, photographs, and newspaper clippings. Carrol Faist, Gifford Miller, Phil Pister, Stewart Udall, Dick Vincent, and Bob Wiley provided firsthand insights into the world of fisheries management from decades ago, along with invaluable perspective on the present day. Jerry Smith, at the University of Michigan Museum of Zoology, provided extraordinary insights, letters, and photographs about the Green River rotenone project. Becky McCue sought me out to share letters, photographs, and stories about her grandfather, Livingston Stone. Many of these she obtained from her father, Edmund Cushing Stone, and from another relative, Edith Perrin. Directly or indirectly, all three of them gave me much valued insights into a man I have come to admire greatly.

Thanks also to those who allowed me to encumber them at work and in the field: to John Riger and Chris Hertrich at the Colorado Division of Wildlife, who allowed me to experience fish culture firsthand and gave me a new perspective on its merits; to Barry Nehring, also of the Colorado Division of

Wildlife, who gave me an insider's history of whirling disease and showed me its effects in the Gunnison Gorge; to Fred Allendorf, who shared his knowledge of trout genetics from his lab at the University of Montana; to Roland Knapp, Esther Cole, and John Ingram, who allowed me to accompany them into the high Sierra during field season; and to Curtis Milliron at the California Department of Fish and Game, who made one such trip possible and tirelessly put up with my repeated requests for interviews and explication.

Robert Behnke provided me with ideas and inspiration throughout the process as well as a careful review at the end. Peter Moyle and an anonymous reviewer also helped improve the book.

I am grateful to Alice Tasman, my agent, who so adroitly steered me and this project through the world of publishing and into the hands of Yale University Press. And I am grateful to Jean Thomson Black and Phillip King, my editors there, who guided this book into reality.

I wrote much of this book during a fellowship at the Center of the American West at the University of Colorado. What a tremendous place to work. The moral, logistical, and financial support of the board, the affiliates, and the staff all played a crucial role in its production, and I don't know what I would have done without Roni Ires to steer me through the logistical details. Likewise, I am most grateful to the National Science Foundation, which funded my fellowship through Award 0522262.

I have also been fortunate enough to have two tremendous mentors as this project has grown from seed to fruit. David Skelly, my Ph.D. adviser at Yale, honed my thinking with his critical analysis even while he gave me the confidence to pursue my ideas to their end. Patty Limerick, director of the Center of the American West, provided inspiration, direction, and unwavering support that kept me going through good times and bad.

To my parents, for the unconditional love, support, and guidance they have given me all my life, and to my children,

Will, Charlie, and Toby, who have taught me the true meaning of joy, thanks. Truly.

And finally, to Ginna—wife, editor, shoulder, rock—this book would not exist without you. Thank you for being the person you are and thank you for being married to me.

A Less Bold and Spirited Nation

I n the year 1872, Lee's surrender at Appomattox and Lincoln's assassination were still fresh in the memory of most Americans. There were only about 40 million people in the United States, about as many as there are just in California today, and women were allowed to vote only in the states of Wyoming and Utah. Telephones and light bulbs were still in the future, horses and carriages still crowded the streets of America's great cities, and President Ulysses S. Grant was fighting a bitter reelection campaign against publishing magnate Horace Greeley. A thirty-three-year-old of modest origins named John D. Rockefeller had just seized control of most of the oil refineries in Ohio, the refining capital of the country. And the completion of the transcontinental railroad three years earlier meant that, for the first time, Americans could travel from New York to San Francisco without sailing around the tip of South America, hazarding the crossing of the Isthmus of Panama, or undertaking an arduous and risky wagon trip across the mostly uncharted plains and great mountain ranges of the West.

On an August day of that year, a train pulled out of Sacramento along the new iron rails, belching a cloud of smoke and steam that was visible from miles away. The engine was not one of the enormous diesels so common today, but a black steam engine so small and quaint by modern standards that it is hard to believe at one time it represented the forefront of industrial technology and power. Over the course of the day, the train chugged through the farms and ranches and occasional settlements in the broad flat valley, the Coast Range low on the horizon through the windows on the left and the Sierra Nevada on the right. Where the two mountain ranges converged and the valley disappeared, a small wooden sign announced that the train had arrived in the town of Red Bluff, California.[1]

The town was not much more than a collection of jumbled wooden buildings and empty lots, but it stood at the end of the line and so the depot was a busy place, full of people and supplies headed to and from the communities that had sprung up in that region during the gold rush. The sound of hammers and the shouts of teamsters must have been incessant, optimistically announcing the birth of another pioneer town and another step toward America's manifest destiny. Ahead, a crew of seven hundred Chinese men labored to cut a grade and lay track into the mountains to the north, while to the south, the wide, flat Central Valley from which the train had come stretched away between the mountains as far as the eye could see.[2]

As the freight from the trains was unloaded into great stacks on the boardwalk or into waiting wagons, the passengers, too, disembarked. And stepping down from the train on that day was a small crew of men, led by one of unremarkable stature who, even in a golden era for facial hair, had a remarkable set of whiskers. Though his chin was clean, he had a mustache connected to sideburns, known as dundrearies, that grew down past his jaw almost to his collar. Weariness, I imagine, must have competed with excitement and anxiety for control of his face, but there was no hesitation.

After querying some of the locals, the men strode over to

the nearby stagecoach office and purchased a ride on the next available coach headed north. The stage road was really just two ruts that were hard and dusty in the dry season and often impassable when it was wet, and the ride, in a claustrophobic carriage with limited suspension and hard wooden wheels bound by iron, could not have been pleasant. Nevertheless, the men endured for fifty miles until the stage halted at the Pit River.[3]

Carriage, horses, and men climbed onto a ferry—a rough wooden platform attached to an overhead cable that stretched from bank to bank—and crossed over. And on the other side, the men gathered their belongings and set out on foot. Following a single-track trail along the bank that had been used by Indians from time immemorial, they came to the mouth of another river known as the McCloud. Here, they turned and followed the trail north along the tributary's banks. They pushed through the manzanita brush, past moss-covered oaks and enormous firs and pines, until, in about two miles, they came to a sandy cove alongside the river. Just beyond, they spied the site of an Indian camp. It was dotted with fire holes and blackened, heat-shattered rocks, apparently the result of generations of use. Ancient graves covered the knoll upstream. And on the nearby bushes, hundreds of fresh salmon had been set out to dry.[4]

It must have been a moment of awe as well as trepidation. For one thing, as their arduous trip must have fully impressed upon them, the McCloud was exceedingly remote. It was also dangerous. The camp the men gazed upon belonged to the Wintu Indians, a tribe that had killed or scared off every miner and settler who had visited the region, except for two miners who had recently gained a tentative foothold high in the headwaters and one silver-tongued soul who had somehow persuaded them to let him marry into the tribe instead.[5]

Many whites disparaged the tribe as "unkempt and squalid," and had no sympathy for these native inhabitants. But a notable minority also believed the outrages of the white men had given the Indians ample cause for their aggressive defense of the area, and I suspect the explorers who looked on the camp

that day in August 1872 probably would have agreed. Nevertheless, tensions between the Wintu and the whites were high, and
the men had every reason to fear for their lives.[6]

As the explorers well knew, though, it was in many ways
because of the Wintu that they were there at all. Other tributaries of the Sacramento had been scarred and degraded by logging and mining to a degree that must have been staggering to
behold. In one commonly used technique known as hydraulic
mining, entire hillsides were blasted away with jets of water.
The tailings from such operations filled entire valleys and muddied the Sacramento River for more than four hundred miles
as well as the magnificent San Francisco Bay. According to one
chronicler, enough sediment had been washed into the Sacramento in one year to raise its bed by more than six feet. Needless
to say, such degradation had disastrous consequences for the inhabitants of the river; fish had virtually disappeared from many
of California's waters.[7]

Thanks to the Wintu's spirited defense, though, the Mc-
Cloud River was still relatively untouched. The river's banks had
never been ripped away by hydraulic mines and the valley still
possessed many huge trees, part of a forest that stretched away
to the north, virtually uninterrupted, until it reached the snow-
covered and solitary summit of Mount Shasta. "It was quite impressive," wrote one member of the expedition sometime later,
"to reflect that we were beyond the white man's boundary, in the
home of the Indians, where the bear, the panther, the deer, and
the Indian had lived for centuries undisturbed."

Yet even though the men must have been agog at their surroundings, it was the fish drying on the bushes that likely caught
their eye the most. For the men were the agents of a new federal
authority known as the United States Fish Commission. And it
was salmon like these that they were seeking.[8]

FLASH BACK EIGHTEEN years, to a valley half a world away.
Picture a unit of British cavalry, checking their weapons and pre-

paring their excited horses to charge. At the head of the valley in which they waited, a place on the Crimean peninsula in the Ukraine that has been known ever since as the Valley of Death, there stood a battery of Russian artillery, ready to send a lethal barrage into anything that should come within range. When the order rang out, the men pricked their spurs into the flanks of their mounts and lunged forward. Within minutes, more than a third of the approximately 670 soldiers were killed or wounded and the charge was turned back. Nevertheless, the courage and the blind devotion to duty displayed by these men awed the world. Alfred, Lord Tennyson, was inspired to write *The Charge of the Light Brigade,* from which come those famous lines, "Theirs not to wonder why / Theirs but to do or die."

What could inspire such valor? This was the question, Tennyson claimed, that all the world wondered. The Russians who viewed the charge initially believed the men of the Light Brigade must be drunk. On another continent, though, a Vermonter by the name of George Perkins Marsh had a different answer. The martial heroism displayed by the British troops, he declared in a report submitted to the Vermont legislature in 1857, was the result of the training their officers had acquired through hunting and fishing.[9]

Three years before the secession of South Carolina initiated the American Civil War, military affairs were on the minds of many in the United States. "The people of New England are suffering, both physically and morally, from a too close and absorbing attention to pecuniary interests, and occupations of mere routine," Marsh wrote in his report. "We have notoriously less physical hardihood and endurance than the generation which preceded our own, our habits are those of less bodily activity; the sports of the field, and the athletic games with which the village green formerly rung upon every military and civil holiday, are now abandoned, and we have become not merely a more thoughtful and earnest, but, it is to be feared, a duller, as well as a more effeminate, and less bold and spirited nation."[10]

The British did not win the Crimean War by being thoughtful and earnest, let alone dull or effeminate. And likewise, Marsh concluded, the only way New Englanders could maintain the rights and liberties they so prized was through courage and self-reliance. The report in which Marsh made these assertions, though, was not a military treatise. It was, instead, a summary of his investigation of Vermont's fisheries. Marsh's thoughts on martial heroism entered the document because he believed the best way to foster such virtues was to encourage sports of the field.[11]

Marsh is best known today as one of the seminal conservationists in the United States. Born to a prominent Vermont family, Marsh forged a remarkable career as a lawyer and a congressman, and ultimately as Abraham Lincoln's ambassador to Italy. He also found time to write numerous books, including a dictionary of the Icelandic language and a biography of the camel. His most well known book today, *Man and Nature*, was first published in 1864 and is still in print. Although it is hard to imagine in this age, when environmental issues make the news every day, this book was one of the first to consider the detrimental effects of civilization on the natural world. And many of the themes in the book were anticipated in Marsh's report to the Vermont legislature, in which he noted that human activities were rapidly diminishing the numbers of salmon, trout, and other fish in New England.

The decline, according to Marsh, was due to a variety of factors. Overharvest probably played some role. Factories and other industrial establishments polluted the rivers with untreated effluent. Dams blocked the migration of fish. The logging industry and the clearing of the land for agriculture had also done their part. Sawmills dumped sawdust into the streams. Denuded hillsides and plowed fields sent tons of sediment into the riverbeds, smothering the eggs of the fish as well as the insects and other invertebrates on which they fed. The temperature of the streams had also risen as the trees that had formerly shaded them disappeared. In consequence, by the mid-nineteenth century, fish that had once abounded had nearly or entirely disappeared from many waters in the state of Vermont. And if the fish disappeared, Marsh believed, so too would the fishermen, along with such things as "dexterity in the arts of pursuit and destruction," courage, and self-reliance. "Nor is there anything in our political condition," he declared, "which justifies the hope, that any other qualities than these will long maintain inviolate our rights and our liberties." Say goodbye to recreational fishing, in other words, and say goodbye to American democracy.

What to do? Like most of his fellow Americans at that time, Marsh believed in progress. Restricting the industry and agriculture that were turning the United States into one of the most powerful countries in the world was not even worth considering. "The unfavorable influences which have been alluded to are, for the most part, of a kind which cannot be removed or controlled," Marsh wrote. "We cannot destroy our dams, or provide artificial water-ways for the migration of fish, which shall fully supply the place of the natural channels; we cannot wholly prevent the discharge of deleterious substances from our industrial establishments into our running waters." The loss of wildlife, in other words, was an inevitable result of civilization and progress.[12]

Neither did Marsh believe in regulating people's rights to hunt and fish. For one thing, any restrictions on these sports would be self-defeating; they would diminish the courage and self-reliance Marsh wanted to inspire. In addition, there were still people in the United States who had been alive during the American Revolution, and cultural memory of that conflict was still strong. Many Americans associated game laws with the aristocracy of the old world, where only the landed gentry had the right to kill many types of fish and wild animals. "The habits of our people are so adverse to the restraints of game-laws, which have been found peculiarly obnoxious in all countries that have adopted them," Marsh wrote, "that any general legislation of this character would probably be found an inadequate safeguard."[13]

Instead, Marsh believed there was an alternative to regulation. Hunting, he believed, would have to be abandoned. In England, horse-mounted gentry were entitled to pursue the game wherever it should run, yielding great sport for the pursuers, but often causing great damage to the farms of the less well off in the process. For America's many small property owners, such a hunt was a nonstarter. More important, the big game animals were rapidly disappearing, and Marsh believed that agricultural and urban development made it unlikely they would ever return. "We must, with respect to our land animals, be content to accept nature in the shorn and crippled condition to which human progress has reduced her," Marsh concluded.

Recreational fishing, though, was another matter. Marsh asserted that industry, which was responsible for the decline, could also provide a technological fix for the fisheries through a wonderful new innovation known as fish culture.

RAISING FISH IN enclosed ponds from which they can be easily harvested appears to be an ancient art. Chinese documents from as early as 2100 B.C. mention laws that regulate the time of year when fish spawn could be collected, suggesting that raising these animals from naturally fertilized eggs was a common practice for many years before that. Artificially raised fish also featured prominently in the sumptuous banquets that occupied a central part of the social scene in the age of the Roman Empire. And during the Middle Ages, the pope somehow reached the conclusion that fish (as well as some of the tastiest waterfowl) were not made of flesh and could therefore be eaten during times of fasting and mortification. Fish ponds, the remains of which can often still be seen today, were established at many monasteries and estates, enabling the owners to eat well on Fridays and during Lent, and also supplying them with a lucrative source of revenue. Fish farming and the transfer of live fish from one body of water to another were thus relatively common practices by the nineteenth century.[14]

Almost all of these efforts, though, relied on natural reproduction; the fish were allowed to spawn on their own, and humans simply collected and raised the fertilized eggs. A French monk in the fifteenth century may have succeeded in artificially propagating fish by squeezing egg and milt together in a pan, but nobody is certain, and in any case his work was forgotten soon after he died. A German army officer also achieved some measure of fame by artificially propagating fish in the eighteenth century, but his efforts were never replicated.[15]

Credit for Marsh's notion of fish culture as a panacea, as a means of restoring the fishes of New England without regulating angler or industry, must go instead to a poor French carpenter and sometime fisherman named Joseph Remy and his friend Antoine Géhin. Fourteen years before Marsh submitted his report, Remy scratched a letter to his local prefect in which he described his efforts to improve the fishing in the headwaters of the Moselle River in France. The technique he described is essentially the same that is used today.

"During spawning season, at the beginning of November, at the moment when the eggs are loose in the belly of the trout," he wrote, "I have, by passing my thumb along and lightly pressing the belly of the female, so that it does not result in any harm for her, forced out eggs that I placed in a pot full of water. Afterward, I took the male, and with a similar operation as for the female, I made the milk run onto the eggs until the water was white. After this operation was completed, and the eggs became clear, I deposited them in tin boxes pierced with thousands of holes and full of large grains of sand. I placed one of these boxes in a running water fountain, and others in the water of the river, La Bresse, in a rather quiet place, though the water was running a little."[16]

Remy subsequently described the hatching of the eggs and claimed that, through this technique, he had populated the river with numerous fish. And although it took several years, the prefect and ultimately the rest of the country eventually recognized the import of the letter. Based on the methods Remy described, in 1852 the French government built a *piscifactoire* (literally a fish

Artificial propagation techniques have changed very little since the nineteenth century. Today, as then, the process begins with a pan, two trout, and some gentle squeezing.

factory) at Huningue, on the border with Switzerland. The one-time poor carpenter and fisherman was feted around the country, and he even had dinner with Louis-Napoléon, president of the Republic. He and Géhin, with whom he perfected the technique, each received a pension and a tobacco shop, the latter apparently a relatively common means of rewarding citizens for their service to the country.[17]

A few years later, in 1853, fish culture came to the United States. An Ohio doctor named Theodatus Garlick read about the efforts then under way in France and decided to try the technique with the native eastern brook trout. Although the scale was small, his efforts were markedly successful, and Garlick was soon raising his artificially propagated fish in ponds on a friend's farm near Cleveland, Ohio. Whether out of his compassion for humanity or because he did not recognize the profits that could be made through a fish-farming monopoly, Garlick made no attempt to keep the technique secret. In fact, he published a book describing just how to go about it.[18]

So it was that Marsh learned of the technique and concluded that fish culture was the solution to the decline of the fisheries of New England and, of course, the virility of its men. And so it was that a crew of fish culturists came to be standing on the McCloud River in northern California in 1872. There's a little bit more to the story, though.

 T W O

Essentially a
National Matter

With the eruption of the Civil War in 1861,
efforts to instill martial values in American men by encouraging
recreational fishing were put on hold. Instead of fishing rods,
men were given guns; instead of animals, men targeted each
other. In addition, with so many of the nation's people and resources dedicated to the war, there was little interest in the still
experimental field of fish culture. But when that bloody conflict
had ground down to its close, the idea of artificially propagating
and growing fish gained renewed interest. Within five years of
Lee's surrender at Appomattox in 1865, at least two hundred
private fish hatcheries had begun operation in the United States,
primarily producing the same fish with which Garlick had initially experimented, the eastern brook trout.[1]

For many, it was a highly profitable enterprise. Including
the cost of building a hatchery and feeding and caring for the
fish, it cost about fifteen to twenty cents to raise a pound of
brook trout. Packed on ice and shipped to Boston or New York

on the rapidly growing rail network, these same fish could be sold to restaurants and fish markets for about seventy-five cents a pound.[2]

In the early years of fish culture, the sale of fertilized fish eggs could also be lucrative. Some eggs were sold to would-be fish culturists who did not yet have mature fish with which to create their own eggs. Some were apparently sold as decorations. An article on the topic in *Harper's Magazine* from 1868 gave precise instructions on raising the eggs and noted that a hatching box would "require no more attention than a globe of gold-fish, far less than an aquarium, afford a far more interesting study than either, and be quite as much of a parlor ornament."

Many of the eggs, though, were sold to landowners who would hatch them and, shortly thereafter, stock the fry into their private waters in the hopes of improving the fishing. In 1872, one fish culturist declared, there was probably not a single stretch of stream in private hands in the Northeast in which fish stocking had not been conducted or at least contemplated. And according to modern historians, the fish culture movement was "the very first environmental crusade," or at least the first crusade to achieve any sort of mass appeal with the American public.[3]

Most of the biggest rivers and lakes in the United States were open to the public, however, and it was in the administrators of those waters that fish culture ultimately found its most important adherents. By 1868 New York and all of the New England states had established state fish commissions, and other states soon followed suit. Although these agencies worked at getting fishways constructed around dams and enforcing fishing laws, much of their effort was dedicated to propagating fish, especially brook trout, black bass, shad, and salmon, and stocking them into public waters.[4]

The only problem was that the pollution, logging, and intensive agriculture that Marsh had described in 1857 had only intensified in the years since that time. And if improving habitat by regulating these industries was infeasible (as many including Marsh believed), then stocking streams with the same fish that had been driven out by the degraded conditions was a stopgap measure at best. "The brook trout must go," wrote one well-known fishing writer. "It is sad to contemplate the extinction of the 'anglers' pride' in public waters, but the stern fact remains that in this utilitarian age its days are numbered and its fate irrevocably sealed. As the red man disappears before the tread of the white men, the 'living arrow' of the mountain streams goes with him."[5]

If they were going to maintain virility and democracy through recreational fishing, then what the anglers, the fish culturists, and the state fish commissions needed was not more eastern brook trout. These fish simply weren't hardy enough. They couldn't tolerate the sediment, pollution, and high temperatures that were becoming the norm in so many streams. At the same time, the native fish that were capable of surviving such conditions, like the catfish, were scorned by anglers who feared that catching such bottom-feeders would endanger their reputations as gentlemen. What these men really needed was a quarry that was both hardy and game and preferably had some claim to aristocracy. What they needed was another trout.[6]

THE FEDERAL GOVERNMENT plays such an overwhelming role in the funding of scientific research these days that it can be hard to conceive of the world in any other way. Researchers (including me), universities, and private corporations are often dependent on the tens of billions of dollars in grants handed out by the federal government every year. Federal agencies conduct their own research and also promulgate regulations affecting every aspect of science and natural resource management. The extent of the federal efforts may be controversial at times, but few people expect the government to abdicate this role altogether. Such, however, was not always the case.

Until the latter half of the nineteenth century, the Smithsonian Institution was the only federal agency that pursued scientific research unrelated to exploration and mapping, and its funds were extremely limited. In 1870, many of today's most important environmental laws were still a hundred years in the future, and even the Progressive Era of Theodore Roosevelt and Gifford Pinchot, the era commonly credited with initiating centralized scientific management of the nation's natural resources, was still three decades away.[7]

So when Spencer Fullerton Baird, the forty-seven-year-old assistant secretary of the Smithsonian, requested funds from Congress to investigate the declining coastal fisheries of Massachusetts and Rhode Island, he was treading new ground. Born in Reading, Pennsylvania, Baird had spent much of his youth and all of his adult life studying the natural history of American wildlife and had become one of the foremost members of that era's scientific establishment. Baird had not attained his influential position solely through his scientific acumen, however. Time and again during the course of his career, he had demonstrated rare political savvy, and this time was no different. He flattered and cajoled congressmen, many of whom he had long ago befriended, and within a few months, he had carried the day. Over the objections of legislators like one from Missouri who presciently declared that if the measure was approved, "there will be no end to the expenditures of public money before we get through with it," Congress passed a resolution authorizing the formation of a United States Fish Commission and provided it with five thousand dollars to get started.[8]

To make it more palatable to congressmembers from other areas of the country, the legislation did not restrict the commission's activities to the New England coast. And as a concession to those who did not believe in a large federal government, the commission was set up as a temporary investigative body, not a permanent new agency. Nevertheless, there was no sunset clause, and when President Ulysses S. Grant signed the bill into law and appointed Baird as the commission's head, he created an entity that would one day evolve into the United States Fish and Wildlife Service.

The Fish Commission, though, had a troubled beginning.

Baird's investigation into the decline of the marine fisheries of Massachusetts and Rhode Island proved both disappointing and frustrating. For one thing, he was forced to acknowledge that his analysis had been wrong. After a summer of research Baird had concluded that the large commercial fish traps that covered much of the coastline of Massachusetts and Rhode Island were responsible for the problem. If they were not removed, he warned, the primary object of his research, a fish known as the scup, would continue to decline and eventually go extinct. Instead, though, a scup baby boom occurred, and the population rebounded.[9]

Baird's proposed remedy, the regulation of the traps, also became a source of frustration. He had initially suggested that the states themselves regulate the traps, requiring the trappers to close their traps on certain days of the week during the spawning season. He reasoned that this measure would allow enough fish to make it to the spawning grounds to perpetuate the fishery. The states, however, did not follow Baird's suggestion. He tried to cajole the legislators with the threat of a federal law that would completely ban all such traps, but by the beginning of 1872, Baird was forced to acknowledge that such federal intervention was unlikely, both because political will was lacking and because the constitutional authority of the federal government in such matters was vague.[10]

At about the same time, a group calling itself the American Fish Culturists Association held its first annual meeting in Albany, New York. Formed a year earlier, the group sought to promote the new fish propagating techniques that were showing such promise. At the Albany meeting, the members of the association decided to push the federal government to begin culturing fish with which to stock public waters, and they formed a committee to forward the project. The members of the committee wasted no time, and shortly after the meeting concluded, a delegation set off for Washington, D.C.[11]

Baird had earlier rebuffed suggestions that the U.S. Fish Commission engage in fish culture. Because of his setbacks, however, Baird was looking for new reasons to justify the existence and continued funding of the U.S. Fish Commission. So when the delegation from the American Fish Culturists Association approached him to suggest his commission take up the cause, Baird replied that he would be happy to undertake the project should Congress provide funds and authorization.[12]

To this end, the culturists turned to Robert Barnwell Roosevelt. Insofar as he is remembered at all, Robert Roosevelt is chiefly known today as Theodore Roosevelt's eccentric uncle. Before his nephew made a name for himself, though, Robert Roosevelt was one of the country's most prominent citizens. Then serving his one and only term as a representative in Congress, the forty-two-year-old New Yorker was a natural ally. An avid angler, he was also one of the first commissioners of the New York Fish Commission. This agency, formed in 1868, had already begun to experiment with culturing and stocking fish in New York's public waters, and Roosevelt was a strong proponent of the practice. Thus, only a month after the members of the American Fish Culturists Association decided to push for federal fish hatcheries, Roosevelt introduced a bill that would appropriate ten thousand dollars to the U.S. Fish Commission to do just that. In his speech on behalf of the bill, Roosevelt reiterated the arguments that had been developed by the fish culturists. Pointedly recalling the difficulties Baird had faced when he suggested regulating the fish traps in Massachusetts and Rhode Island, Roosevelt proclaimed that fish culture was indisputably the province of the federal government.

"It is," he declared, "essentially a national matter; the States alone cannot take charge of it and manage it efficiently; they cannot even pass laws which will thoroughly protect the fish at seasons when they should not be disturbed. Rivers run through different States, or are the boundaries between them, and the laws made for part or for one shore might not be identical with those made for other places. Unity of action is essential, for it is useless to protect in one locality if wanton destruction is permitted in another."[13]

As for the technique itself, Roosevelt reported that "it is not an untried theory, resting more in hope than experience, but has passed from the realm of experiment into absolute certainty." Roosevelt noted that although an individual fish may produce thousands of eggs, only a few ever make it to maturity, the rest being destroyed by disease, predation, and other dangers. If it reduced these risks by spawning and protecting the eggs and young fish in a hatchery, he declared, the government could increase the population many times over. Roosevelt's hometown paper was even more optimistic. An article in the *New York Times* written at almost exactly the same time noted that "the expenditures of a few hundreds or thousands of dollars annually in any river will tend to throw such an enormous number of young fish into the water as it would be quite impossible to catch all the matured fish by any means tried."[14]

Thus, in an interesting combination of philosophies that would have found many adherents in an age when faith in technological progress rivaled faith in God, and Transcendentalism had raised the worship of nature to a similar plane, Roosevelt laid out his vision of the future.

"Fish are exceedingly prolific; nature seems to have made them the great store-house of food which was to be held in reserve until an increasing population should have required it for support," he declared. "Every need of the human kind seems to be met as it is developed, and the earth apparently holds in its recesses the secrets which are to keep the world thriving and progressing for ages, and until it shall be covered with a swarming and happy population, denser than is now imagined possible, or than learned essayists on a subject they do not comprehend would permit as at all prudent. Fish food is manifestly one of the means which are to make such a result possible."[15]

Opponents blocked Roosevelt's bill on a rule of order. However, a virtually identical bill appropriating fifteen thousand dollars to the U.S. Fish Commission to commence fish culture did eventually pass the House and Senate, becoming law on June 10, 1872.[16]

The powerful and charismatic Spencer Fullerton Baird led the U.S. Fish Commission in its early years.

Smithsonian Institution Archives, RU 95(2)

Three days later, Baird held a meeting in Boston with state fish commissioners and members of the American Fish Culturists Association. They focused on two fishes, both of which were mentioned by name in the appropriations bill. The first was shad. Under the terms of the bill, the fish commissioner was required to introduce this fish into the waters of the Gulf states and the Mississippi Valley. Baird had suggested this provision be included in the bill to gain support from the congressmembers from these states and because he thought the introduction of this species was likely to be successful and beneficial to the citizens who lived near these waters.[17]

The second was salmon. Atlantic salmon had been eliminated or sharply reduced in the rivers of the eastern United States, wiping out what had once been a valuable fishery. Everybody at the meeting hoped to rectify this problem through fish culture. The question was where to obtain the eggs. Canada was out. Relations between the two countries were at a low ebb at that time, and Canada jealously guarded its supply of these fish. Another option was to try to obtain the eggs from Maine or Germany, but most of the men at the meeting agreed that neither location could provide a sufficient supply.[18]

The third option was to try to obtain eggs from the salmon of the Pacific. These fish were thought to be more prolific than their Atlantic cousins and thus, it was believed, they would be more than able to supply enough eggs. The Pacific salmon would of course be a different species from the Atlantic salmon, but this was not, apparently, a cause for concern. The introduction of new species was, after all, part of the mission of the U.S. Fish Commission.

The only worry, at least from Baird's perspective, was whether the Pacific salmon would strike an angler's fly. Indeed, there was a story, probably apocryphal, but nevertheless popular among certain circles in the 1870s, concerning just this question. According to this myth, the British had ceded the Oregon Territory to the United States in 1846 because the brother of Prime Minister Lord Aberdeen had been skunked during a fly-fishing expedition on the Columbia River. The man reportedly declared to his brother that the region was not worth fighting over because the salmon would not take a fly.[19]

Baird's concerns were telling, because even though he publicly justified his agency's activities for utilitarian reasons— claiming fish culture would supply the country with a cheap source of food—he clearly recognized that sportfishermen like Robert Roosevelt and many other members of Congress were already an important part of his constituency.[20]

Baird also objected to the Pacific salmon because he had heard that many of them died after spawning, unlike their Atlantic cousins. Ultimately, however, he relented and agreed to the plan. In many ways, it was probably a foregone conclusion. To curry political favor, Baird had tacitly agreed to establish a West Coast hatchery when Congress was considering the enabling legislation. And so, Baird turned to the meeting's biggest

proponent of a Pacific salmon hatchery, a thirty-five-year-old
New Hampshire man named Livingston Stone, and asked him
to initiate the project.[21]

LIVINGSTON STONE WAS born in Cambridge, Massachusetts, in 1836 to a family whose roots in that town extended back
to the year it was incorporated as part of the Massachusetts Bay
Colony two centuries earlier. After graduating from Harvard,
he attended theological school in Pennsylvania and, in 1864, became the pastor of a Unitarian Church in Charlestown, New
Hampshire, a bustling town on the Connecticut River a hundred
miles from Cambridge. His unimpeachable and relatively unremarkable curriculum vitae, however, appears to have hidden a
more complex character. When he accepted the job in Charlestown, Stone wrote a letter that would give any employer pause.
"I have to say that after a serious consideration of the subject
which at first presented many conflicting aspects, I have decided
to accept the invitation extended to me, and to take my lot for
better or for worse with this society," he wrote. "And let me add,
that having now decided to put my hand to the plough there
shall be, on my part, no looking back."[22]

      In 1868, however, he resigned his post, declaring that "for
some time past, owing to various causes, it has seemed to me
that is [sic] was not for the best interests of this society, that I
should retain any longer my connection with it as Pastor." Sen-
tences like these are enough to make anybody wonder if there
was something hidden in the closet, and for some time while I
was researching this book, I became obsessed with this mystery.[23]

      In later years, his biographers would claim that "failing
health compelled him to seek an outdoor life." And indeed,
Stone lived in an age when tuberculosis and other debilitating
diseases were rampant, and the most commonly prescribed
cure was to spend as much time as possible in the sun. So accepted was this method of treatment in the era before antibiotics

that the journal of the National Tuberculosis Association in the United States was known as the *Journal of the Outdoor Life*. But Stone never mentioned his health in his letters from the period, and he did not do so in his letter of resignation, despite the fact that it would have offered a relatively easy out.[24]

Neither does it appear that he had any moral shortcomings or spiritual misgivings. In fact, upon accepting his resignation, the members of the church passed resolutions that were "very complimentary to him, both as a preacher and a man," and that "extended to him the assurance that he would carry with him the best wishes of the society for his future happiness and prosperity." These do not appear to have been empty promises. Stone retained his connection with Charlestown for many years, and eventually he even married the daughter one of the most prominent members of the church, who was at one time a New Hampshire Supreme Court justice.[25]

So what caused him to abandon such a safe and honorable career? I have visited Charlestown. I have stood in the church where Stone used to preach. I have seen his house, and the town hall in which he first danced with his future wife. I have squinted at pictures and tried to imagine what it must have been like, standing in his shoes more than a hundred years ago. The wonderful caretakers of the Charlestown Historical Society have showed me letters and artifacts from Stone and his contemporaries. And at one point, Stone's granddaughter contacted me, out of the blue, and shared more of his letters and her recollections.

No smoking gun ever emerged, though. Only a portrait of a caring and responsible man who had a touch of wanderlust in his soul. Perhaps something else will turn up one day. For now, I can only conclude that Stone departed the pulpit because he was a searcher, the type of man who could not content himself on a road that had been traveled so many times before.

For a couple of years before he resigned from the ministry, Stone had been experimentally propagating and raising eastern brook trout, setting up fish ponds and a hatchery known as the

Cold Spring Trout Ponds only a few hundred yards from his church in Charlestown. He found it thrilling. "The present age of almost daily recurring marvels had hardly begun then, and people were more incredulous and slower to accept apparent miracles than they are now," wrote Stone many years later. "It seems that we should never feel again, and we probably never shall, the thrill of pleasing excitement that tingled to our finger ends when we first saw the little black speck in the unhatched embryo which told us that our egg was alive."[26]

Stone found he was spending more and more time at the ponds, and in 1868 he simply decided to dedicate himself full time to his newfound vocation. As the first fish farm in New England, it was for some years a very lucrative business. Stone believed he was at the forefront of a fantastic new human enterprise that many people believed would one day outstrip conventional agriculture in terms of the amount of food it could produce on a per-acre basis. He sold his fish in the cities and even offered a kit with one hundred fish eggs and a hatching trough that was suitable for use as a home decoration, an item he advertised as a "novel Christmas present."[27]

Stone never seems to have been overly concerned, however, with amassing great wealth. Instead, he appears to have pursued fish culture for the same reasons he initially pursued the ministry; he possessed a keen sense of compassion and humanity. Instead of limiting competition by keeping his techniques a secret, as did other culturists, Stone was more than happy to dispense advice to others who were interested in beginning their own operations. At one point, Stone wrote a column in a newspaper on the best means of raising fish, and he later wrote a book on the topic. Stone was one of the founding members of the American Fish Culturists Association, and a member of the delegation that persuaded Roosevelt and Baird to initiate a federal fish culture program.[28]

Eventually, though, Stone's open nature became his downfall. As more and more fish farms opened, many of them following his advice, Stone's profits fell, and within a few short years, he was struggling financially. At one point, he was forced to take his hat in hand and beg his nephew for money, embarrassedly asking him to "keep it wholly to yourself, till you are made good again." And so, in 1872, when Baird formally offered him a job with the federal government, a job setting up a hatchery in California that both men hoped would one day replenish the salmon fisheries of the East, Stone was quick to agree.[29]

Stone put his cherished trout farm up for sale, gathered a few of his belongings, and purchased a train ticket for San Francisco. On August 1, 1872, he stepped aboard in Boston and, for more than a week, the former minister of the South Parish Church in Charlestown, New Hampshire, traveled across a continent that few people had ever seen. It was a continent where bison herds still roamed the plains in great numbers, a continent that was home to untold numbers of Indians who were still far from being subdued.[30]

After a steam engine belching black smoke pulled him across the Rocky Mountains, through the Great Basin, and over the Sierra Nevada, Stone arrived in the boomtown of San Francisco. That city, too, must have amazed a man who had never traveled far from New England. It was a city that had not even existed in 1846. But with more than 150,000 people scrambling, hustling, and building, it had already grown into the tenth largest metropolis in the country, and it was still expanding. By the turn of the century, San Francisco would hold nearly five times as many people as it did when Stone first arrived.[31]

Apparently undaunted, Stone stepped down from the train and almost immediately began asking questions about where he could find a good run of salmon. He queried members of the California Fish Commission and other fish culturists. However, "singular as it seems," he reported, "I could find no one in San Francisco who was able to say either where or when the salmon of the Sacramento spawned." Ultimately, an engineer from the Central Pacific Railroad showed him a spot on the map where "he assured me he had seen Indians spearing salmon in the fall on their spawning beds." The spot he indicated was on the Mc-

Livingston Stone (center) recruited his nephew, Willard Perrin (right), and Vermonter Myron Green (left) to help build and run the hatchery on the McCloud River. They posed for this photograph in San Francisco in 1873.

Courtesy of Rebecca Cushing Stone McCue

Cloud River, at the top of the Sacramento River valley, fifty miles north of the railroad terminus at the town of Red Bluff.[32]

WHEN STONE AND HIS MEN finally came across the Indian fishing camp on the McCloud and saw the cleaned fish drying in the sun, it was already close to the end of the salmon spawning season. Having come so far, however, the men were not about to give up. Despite the fact that they were "twenty-five miles from the nearest town or village, fifty miles from a railway station, over fifty miles from an available saw-mill, and in the Sierra Nevada Mountains, where the mule-teams barely made twenty miles a day with supplies," they set to work building a hatchery about a mile above the site of the Indians' camp.[33]

Their activities didn't please the Wintu, and they made their feelings clear to Stone. "When we came to the river to erect our house and hatching works, a large number of Indians as-

Kloochy, a Wintu Indian,
was eighty years old
when he posed for this
photograph in 1882.

National Archives,
College Park, RG 22
FFB B508

sembled on the opposite bank and spent the whole afternoon
endeavoring by threats and furious gesticulations to drive us
away, and afterwards several of them waited on me and told me
in their dialect of which I had learned a little, that this was their
river and their land, and these were their salmon, and that I was
stealing the land and salmon; that they had never stolen anything from the white man nor taken his land; and that I ought to
go away. Some of them went so far as to give out threats about
my being killed. When I thought of the fate of all my predeces-
sors on the McCloud, I did sometimes feel slight misgivings,
but I adopted a firm and conciliatory policy with them which
worked so satisfactorily that I am now perfectly satisfied that
none of us are in any danger there."[34]

"Slight misgivings" surely understates the emotions Stone and his companions felt at the moment of crisis, but they did persevere. In a little more than two weeks, they constructed a house, tanks to hold the spawning fish, boxes in which to hold the eggs, and a flume to supply them with fresh water.

As soon as the facilities were completed, Stone and his men set to work trying to catch and propagate them. But despite their Herculean efforts—the men worked day and night netting fish—they only managed to capture twenty-six salmon that had not already spawned, enough to yield but one bucket of eggs.

Shipped east via Wells, Fargo, and Company, the fry from these ova were stocked in the Susquehanna River and never seen again. Optimistic about the salmon hatchery's prospects in future years, though, Stone viewed the operation as a success. In truth, however, the 1872 expedition to California is more notable not for this small harvest of salmon ova, but because it was the first time Stone set eyes on another species of fish, a few of which the men caught and struggled to identify. It was a fish that the resident Indians called *syóolott,* and which Stone referred to as Sacramento River trout, common mountain-trout, Red-banded trout, and simply, trout,—all different names for what is today considered to be one fish: the rainbow trout.[35]

 T H R E E

Let the Best Fish Win

Livingston Stone and his associates on the Mc-
Cloud often get credit, at least among those who care enough to
have an opinion, for being the first to propagate and disperse
rainbow trout outside their native range. It would certainly
make for a cleaner story. But unfortunately, it's not true.[1]

In their early years on the McCloud, Stone and his crew frequently did, in fact, encounter rainbow trout in their nets. They
even collected a few as museum specimens. But their primary
mission was to propagate salmon, and they stayed focused. So
if you are ever playing a fisheries trivia game, you must remem-
ber a group with a mouthful of a name, the Ornithological and
Piscatorial Acclimatizing Society of California, whose mission
was to import "all the game birds and fish of the older states and
Europe." This organization gets the credit for being first.[2]

Formed by sportsmen and prominent citizens of the
Golden State in 1870, the society was part of a worldwide acclimatization movement that, like fish culture, had first blossomed
in France. In many ways, the movement was a direct result of
the French and British quest for empire and the colonization
of distant lands with vastly different climates, flora, and fauna.

Beginning in the sixteenth century, explorers from the colonies and other far-flung locales sent home all manner of specimens that they deemed beautiful, useful, or exotic.

And as explorers became settlers, traffic also began to flow in the other direction. Initially, the colonizers focused on the importation of utilitarian and domesticated species; cows and sheep found new homes in outposts around the world. But as the colonists became more established, they also sought to re-make the natural landscape around them. Across the globe in the middle of the nineteenth century, people formed organizations for the purpose of importing and releasing plants and animals into the wild, usually species that had a strong association with the home country.[3]

The first acclimatizing group, the Société Zoologique d'Acclimatization, formed in Paris in 1854. Members of this group and adherents to its philosophy tended to believe that plants and animals, including humans, could adapt to new climes through physiological changes. They conceived of acclimatization similarly to the way the term is used today when, for example, people adjust to a different altitude. And because they believed that most animals could be forced to evolve and adapt to widely different climates if the transfer was handled properly, the French eagerly transported exotic species throughout their empire. Ostriches and Chinese yams, tapirs and zebras—nothing was too bizarre to at least merit consideration.[4]

An English acclimatizing group, formed in London in 1859, also sought to transfer various species throughout the British Empire, and other groups arose in other corners of the world shortly thereafter. British acclimatizers tended to be less san-guine than the French about their ability to force organisms to adapt to different regions. Instead, they chose to focus their efforts on moving mostly game animals between areas with similar climates. Experiments with fish proved especially fruit-ful. A shipment of brown trout, for example, was sent to Aus-tralia and New Zealand in 1864, and by 1880 these fish had be-come well established on that far continent. In the United States, the movement also excited broad interest. In some cases, the introductions were conducted by groups organized specifically for that purpose, like the Cincinnati Acclimatization Society and the American Acclimatization Society in New York. In other cases, the introductions were haphazard efforts by one or two individuals.[5]

Many American acclimatizers focused on fish. While some were busy importing fish from Europe, like the brown trout and the carp, others dedicated themselves to "distributing the best of our own fishes through all the waters of the continent." Birds were also popular. Some, like the house sparrow, were imported to control pests. Others were brought in for other reasons. One eccentric New York City drug manufacturer named Eugene Scheiffelin apparently concluded that his fellow citizens would be culturally impoverished unless they were exposed to all the birds mentioned by Shakespeare. He diligently set about rectifying the situation. And so today, because they are mentioned in a line in *Henry IV,* the North American continent is now inundated with starlings that Scheiffelin first successfully imported from England in 1890.[6]

As a philosophy, acclimatization fit well with American ideas of progress and manifest destiny. The white man would rightfully and inevitably replace the native people of the continent, civilization would supplant wilderness, and new plants and animals would ultimately oust their native counterparts. "Our only object can be to improve our fishing, and make our stock of sporting fish, if possible, the best in the world," wrote one avid promoter of the idea. "Let the best fish, like the best man, win."[7]

The acclimatization philosophy found particularly fertile ground in the American West, where many believed the fish fauna was somewhat lacking. And there is some truth to the idea, at least in terms of numbers. Even today, biologists have found only about thirty fish species that are native to the San Joaquin and Sacramento River systems in California. Only about half a dozen more are native to the entire state of Arizona. In the eastern states, in comparison, a single creek or even riffle may hold an equivalent number of species.[8]

But it wasn't just the number of species. Most of the fish swimming the western waters were worthless for commercial or recreational purposes, at least as the anglers of the day viewed them. Of the West, Robert Roosevelt declared, "The markets are almost bare of fish; a few catfish, suckers, and pickerel constitute the wretched and meager bill of fare they offer. The muddy Mississippi contains little or nothing. The beautiful Ohio has but one or two sorts of pike-perch, which the inhabitants flatteringly call salmon, while catfish hide in most of the discolored streams of our continent and suckers explore the bottom for their food."[9]

For those who believed in acclimatization (and most people did), the course of action was obvious. "There is no reason why the waters of the West should be less prolific than those of the East," Roosevelt declared, "provided the right species were introduced; and were trout, salmon, bass, shad, and sturgeon to take the place of catfish, pickerel, and suckers, the gain would be manifest."[10]

Consequently, in 1870, the members of the California Acclimatizing Society decided to take matters into their own hands. Most of the members of the society were originally from the East. Their journey to the Pacific Coast had been arduous. Some traveled overland, on foot, horse, or wagon. Others traveled by ship, with some of them sailing all the way around Cape Horn and others crossing the Isthmus of Panama. In other words, just the fact that they were living in California said something about their character. They were the type of people who got things done—the type of people for whom transforming the biota of an area as vast as California seemed like a feasible objective.

The first fish the California Acclimatizing Society sought to import was the eastern brook trout. Although they were not the most hardy, they were still the most commonly cultured fish in the United States at the time. They also probably appealed to many California sportsmen for nostalgic reasons; many of these men had grown up in the eastern part of the country, the native home of these beautiful fish. And eastern brook trout also had a certain cachet. They were the trout most commonly described in the fishing literature of the day and they were associated with the gentlemen of the East, a sophisticated culture that many of the wealthier Californians desperately sought to emulate. They were, according to one San Francisco newspaper, "said to be the best trout known to the American sportsman."[11]

WHEN IT DECIDED TO import eastern brook trout, the California Acclimatizing Society turned to a man named Seth Green. In his pictures from the 1870s, Green looks like one of the patriarchs of the Old Testament, an appearance that seems to have been all the more vivid to those who met him in person. One fawning contemporary asked his readers to "imagine a strong and stalwart frame surmounted by a head strongly resembling that of Socrates, and covered with a white silky beard and luxuriant gray hair." Green was known, even during his lifetime, as "the father of fish culture in America." In fact, if it had been up to Green, the last two words probably would have been dropped. Green insisted that he had conceived of artificially propagating fish in 1838, a full five years before Remy had demonstrated the technique in France. That he did not actually try it for more than twenty years was, he insisted, only because he had been too busy with other things.[12]

Despite (or perhaps because of) his skillfully orchestrated self-promotion, there is no question that Green was a man of tremendous accomplishment. Born in 1817 to a tavern owner in upstate New York, Green was, by all accounts, a tremendous sportsman with both rifle and rod. When he turned eighteen, he went into business as a commercial fisherman and merchant, and within a few years he had achieved no small degree of prosperity.[13]

With the money he had accumulated, Green purchased the land surrounding a productive trout stream in Caledonia, New York, in 1864 and began experimenting with different methods of artificially spawning brook trout. His success was such that when fish culture began to burgeon toward the end of the decade, he operated the biggest hatchery in the country. So when he received the request from the California Acclimatizing Society in the spring of 1871, Green carefully packed ten thousand eastern brook trout eggs and put them on the train to San Francisco. Of greater import for this story, though, is the fact that this shipment initiated an apparently cordial relationship between one of the most influential fish culturists of the day and a group of people who were propagating fish in the heart of the rainbow trout's native range.[14]

It is difficult to know exactly when the California Acclimatizing Society began culturing rainbow trout. Flip through the newspapers from the day and you find that while the dates, numbers, and locations of the society's eastern brook trout efforts are meticulously recorded, rainbows are only casually mentioned in frustratingly vague asides. An article from 1871 notes simply that, "Several of our wealthy and well-known citizens have attempted the cultivation and breeding of our mountain trout, but from various causes, not involving the principle, were failures, or nearly so." The article then goes on to discuss the success the acclimatizing society had with eastern brook trout and with cutthroats from Lake Tahoe. The same newspaper mentions these two species of trout again in 1873, when it states that the society obtained eighty-six thousand eggs from California trout, most of them from the San Andreas Reservoir outside San Francisco. Other reports suggest that the society was successfully breeding these fish by 1872.[15]

Maybe, though, the difficulty in ascertaining the exact date when rainbow trout were first cultured or where they came from is telling in and of itself. The fish that one day would become the most commonly cultured trout in the world seems to have begun its ascent in a strikingly unremarkable manner. Fixated as its members were on importing new species, it seems likely that the acclimatizing society simply began culturing rainbow trout because it was too hard to ignore them.

Although the society built its first hatchery in a building on the corner of Fulton and Gough Streets in the middle of San Francisco, their operations quickly outgrew this facility, and by 1871 the group had established a new location on the San Pedro Ranch, about fifteen miles south of San Francisco. The water in the fish ponds at San Pedro came from San Pedro Brook, which also served as the spawning grounds for native freshwater-resident rainbows and steelhead. The nearby San Andreas Reservoir (which had been recently built to supply the city of San Francisco with water) probably also held large quantities of these fish. With the ponds and hatching facilities already in place, it wouldn't have required much effort for the members of the society to begin culturing some of these trout, which means that rainbows from San Pedro Brook or San Andreas Reservoir were probably the first rainbows ever to be artificially propagated.[16]

Whatever the case may be, in the spring of 1875, the California Acclimatizing Society packed five hundred of these eggs on moss and put them on a train to New York—the first time rainbow trout had been shipped out of their native range. The eggs were apparently a gift to Green as well as the state of New York, for by that time Green had sold his hatchery to the New York Fish Commission and become the superintendent of the state's fish culture operations.[17]

NOT ALL OF THE rainbow trout that were originally shipped to New York survived. Those that did, however, thrived. In the spring of 1878 Green was caring for 275 of these fish. They weighed about a pound each, considerably more than any eastern brook trout at the facility at the same age, or even twice that age. In a subsequent report to the legislature, Green and the New York Fish Commissioners were practically bursting with enthusiasm for their California trout.

"We believe we shall have conferred the greatest possible boon upon the anglers, not only of New York, but of all the

Atlantic States, by the acclimatization of those fish," they wrote. "They are not quite as beautiful as the native trout, wanting as they do the carmine specks, but are not much inferior in appearance, and in the matter of flavor and taste, are fully their equals. But for hardiness, for certainty in hatching and safety in raising, they are far superior, and in game qualities more than their compeers. So vigorous are they that a person accustomed to stripping trout will at first have difficulty in handling them. They take the fly as readily as [eastern brook] trout, and make a better fight against capture."[18]

The report noted that the California trout could also withstand much higher temperatures than the native trout of New York. This was a valuable trait in streams and ponds where excessive logging had turned once cool forest streams into sun-baked wastelands. The commissioners concluded that, with the California trout, "we shall give to the private trout breeder and sporting angler exactly the fish he wants." And indeed, so impressed were these New Yorkers with their new trout, they later decided to try out another variety from California. In May of 1878, Seth Green's son Chester returned from a trip to California with 113 trout fry from the McCloud, a river already made famous by Stone.[19]

 F O U R

As Many Different
States as Possible

B y 1879, Livingston Stone and his assistants on the McCloud had produced 45 million Chinook salmon eggs. And thanks to the growing network of railroads and the distribution system established by the U.S. Fish Commission, these eggs had been distributed to twenty-nine different states, from South Carolina to Maine, and even to interior waterbodies in Colorado and Utah. Yet despite these efforts, "In no single case did the experiment prove satisfactory." Not a single new run of these salmon had become established. It was, Stone conceded, "a stupendous surprise and disappointment."[1]

And so, one July morning in 1879, Stone and two assistants climbed on to horses and set out to find a suitable location for a new hatchery, not for salmon, but for rainbow trout. The country above the salmon-breeding station was still nearly untouched by white settlers, so they followed an Indian trail upstream, "over the cliffs and around the hills of the cañon, among some of the most magnificent landscapes in the world," Stone reported. "We were soon inclosed in a circle of almost inaccessible mountains, where high, precipitous cliffs extended down sometimes to the water's edge, and through which the McCloud somehow wound its tortuous way, not so much a river here as a succession of foaming cascades."[2]

Despite having lived in the area off and on since 1872, Stone clearly was still awed by the scenery. And since he lived in an era when rhetoric was highly valued, when scientists and bureaucrats did not feel bound to muzzle their emotions in their official correspondence, his writing can be a pleasure to read. In dispatches that were published in the annual reports of the United States Fish Commission, Stone created evocative images of the West, literary versions of the landscape paintings created by his contemporaries Albert Bierstadt and Thomas Moran. Perhaps he sought to generate support for his operation with audiences in the East, perhaps he was thinking of posterity. Whatever his motivation, the McCloud had obviously touched him deeply. "The Yosemite Valley is sublime and stupendous in its grandeur," he wrote, "but there is a brilliancy and enchantment about the beauty of the Upper McCloud that I have never seen in the Yosemite or anywhere else."[3]

Ultimately, Stone decided to build the trout-breeding station about four miles above the salmon hatchery on a tributary named Crook's Creek. It was a bold choice, given that the creek was named after a settler who had been killed on that spot by the Wintu Indians in 1873. Nevertheless, Stone and his men set to work building a new facility at the site, including a dwelling house, a hatching house, and holding ponds for the breeding fish. The men also cut poles and heavy timbers from the surrounding hills and built dams, racks, and traps in two of the nearby creeks to catch trout as they ascended to spawn. Within a few months, Stone was pleased to report to Baird, they had gathered "the finest collection of trout ever brought together in one place."[4]

IN ITS FIRST EIGHT YEARS, Baird had transformed the
United States Fish Commission from a small, temporary investigative body to a large quasi-permanent agency. In an era of
laissez-faire capitalism, an era of widespread distrust of the
federal government, this was no small achievement. Baird had
apparently learned his lesson from the way Congress had re-
pudiated his attempts to regulate the scup fishery in 1872. In
the following years, Baird pursued a philosophy that was much
more popular with the public: stocking so many fish that regulation would be unnecessary.[5]

For this reason, the U.S. Fish Commission began supplying
eggs to state fish commissions and private individuals free of
charge. A few days before the eggs were due to arrive, the U.S.
Fish Commission would send the applicant a telegram. All the
recipients had to do was arrive at the train station at the speci-
fied time with some buckets or milk containers into which the
eggs could be transferred. Baird almost certainly believed this
program provided a valuable national service. In most cases,
the Fish Commission shipped the eggs to state authorities or
sporting clubs that diligently stocked the fish in waters that they
believed to be suitable for the species in question, waters that
were usually open to the public.[6]

However, Baird also clearly viewed this program as an invaluable means of currying favor with those in charge of the
federal purse. Letters to Baird from congressmembers, who
were requesting fish eggs for their constituents, fill several boxes
at the National Archives in Washington, D.C. Few of these requests were denied, since it was frequently Baird who had instructed the applicants to route their entreaties through their
congressional delegation. In one letter, Baird was quite explicit
about his aims. "It does not make much difference what Rock-
wood does with the salmon eggs," he once declared in a letter to
a subordinate who was concerned about the survival of the fish.
"The object is to introduce them into as many different states as
possible and to have credit with Congress accordingly. If they
are there, they are there, we can so swear, and that is the end of

From 1881 to 1947, the U.S. Fish Commission delivered fish to private as well as public applicants throughout the United States on specially designed train cars like this one. The cars had water tanks, aeration devices, cooling systems, and bunks for the attendants, among other things.

National Archives, College Park, RG 22 FFB B1001

it." Baird was, in short, a shrewd political operator, and by 1879 he had persuaded Congress to increase the appropriation for his fledgling agency to seventy thousand dollars a year. Between 75 and 85 percent of these funds were spent propagating and stocking fish.[7]

Just what prompted Baird to instruct Stone to begin culturing rainbow trout on the McCloud is not entirely clear. Much of it, I suspect, derived from the fact that both men had invested so much in the McCloud. It wasn't just monetary. Their personal reputations and the reputation of the agency were on the line as well, and neither Stone nor Baird was about to admit failure and give up.

Nevertheless, it was a curious decision. Baird had long resisted propagating eastern brook trout because he believed they were primarily a sport fish and luxury item, not a utilitarian food species. On the other hand, Baird also recognized the tre-

In the nineteenth century, the U.S. Fish Commission raised fish in ponds on what is now the National Mall. Here they are arriving in milk cans on a horse-drawn cart, with the unfinished Washington Monument rising in the background.

National Archives, College Park, RG 22 FFB B539

mendous political clout of the sportfishermen and understood that it would benefit the commission if it were to propagate a fish that was popular with anglers.[8]

Rainbow trout may also have had a leg up on the brook trout as far as Baird was concerned because they were not native to the East. The primary mission of the commission at the time was to acclimatize fish to areas where they did not already exist, not to supplement existing stocks. And while many people were raising brook trout, there were few commercial or state hatcheries growing these western fish.

In addition, Baird had no doubt heard the praise that Robert Roosevelt and Seth Green were heaping on these fish. Not only had the New York Fish Commission been propagating rainbows from the San Francisco Bay area for many years, they had also been propagating McCloud River rainbows, sent to them by one of Stone's employees, a man who had been culturing them in the off-season along with another settler on the McCloud.[9] They reported that these fish were hardy enough to withstand some degree of pollution (think less regulation of industry) and had a high enough temperature tolerance to be stocked across many regions of the country (think lots of congressional districts).[10]

With a fish so praised, a hatchery already in operation nearby, and a man on the payroll who had already propagated these trout, Baird's decision to establish a new facility for propagating McCloud River rainbows was probably close to a no-brainer. But for anglers and aquatic ecosystems around the world, it was also one of the most momentous decisions ever made.

IN ITS FIRST YEAR OF operation, Stone's trout hatchery on the McCloud River cultured more than a quarter million eggs. By 1887, this number had increased to more than 3 million. Obtaining that many eggs was no easy feat. Mountain lions were abundant and sometimes quite bold. One cat sprang out of the woods directly behind a member of the crew who was walking home, and carried away a dog that was following close behind. One man was bitten by a rattlesnake, another by a tarantula, and the scorpions were reportedly thick.[11]

The Indians also posed a problem. Stone had sought to maintain good relations with the Wintu from the beginning and had been one of their foremost champions in Washington. At one point, Stone urged the president to set the area aside as a reservation for the Indians and the fish. On a more personal and local level, Stone did small things for the Wintu, like giving them the salmon he caught after they had spawned. Perhaps most important, he treated them with respect, believing them in many cases to be more honest and upstanding than the miners and settlers who were still flocking to the state. "I would trust the McCloud Indians with anything," Stone later wrote. "I wish on these accounts to be very emphatic in saying the charges against these Indians being a race of thieves, are untrue and unjust."[12]

For at least a few of the Wintu, the feelings were mutual. In fact, some of the Wintu chose to work at the hatchery, throwing themselves wholeheartedly into the job. They were "steady, industrious, and very intelligent," Stone later declared, and particularly good at handling the fish. "If we could not have the Indians to help us, it would be very difficult to supply their place."[13]

But not all of the Wintu were so accepting. At least a few of them understandably seemed to resent the intruders. For whatever reason, they had kept their hostility in check since Stone arrived. But when they learned that Stone and his assistants planned to take over Crook's Creek, their true feelings came to the surface.

A few days after they began building the hatchery, one of the assistants was left alone at the site. According to Stone, "As the day wore on he thought he heard a slight noise near him, and on looking up he saw to his great surprise three Indians standing over him, each with a drawn knife in one hand and a rifle in the other, and here, on the very spot where the last settler was murdered, they told him the same story that they had told the murdered man, viz, that this was their land, that the white men had no business there, and that they did not want white men on the McCloud River at all."[14]

Fortunately probably for all concerned, the Indians chose not to kill the man—an act that would undoubtedly have brought reprisals—and instead decided to let him go after three tense hours. Nevertheless, Stone and his men felt the need to keep arms and ammunition handy for many years. They feared Indian "uprisings," as had occurred among nearby tribes, as well as the potential harm that could come from a few individual Indians acting with or without the support of the other Wintu.[15]

They also feared the outlaws who increasingly made the area home. Stone reported in 1877 that his butcher had been shot, along with one of his neighbors, and the stagecoach had been robbed so many times that an armed guard had to accompany it when it made trips from the hatchery into the nearby town of Redding. "With tarantulas, scorpions, rattlesnakes, Indians, panthers and threats of murder our course here is not wholly over a path of roses," Stone offhandedly remarked in one of his reports.[16]

The fish also faced threats. In the winter of 1881, torrential rain caused the McCloud to rise to a level twenty-six feet higher than its normal summer level. The flood washed away all of the main buildings at the salmon-breeding facility. The trout-breeding station was at a higher elevation and was spared from the flood, but the rain liquefied the hillsides, sending enormous mudslides through the trout-breeding ponds that killed many of the captive fish.

Transportation posed other problems. If the fish were shipped as adults or at any point after they had hatched from the egg, they had to be kept in well-oxygenated water and within a limited temperature range. In an age before modern refrigeration techniques were widespread, maintaining these conditions was no easy task. Whoever accompanied the fish had to constantly change the water and add ice chips to keep the mercury in the thermometers from getting too high. Regarding one sleepless weeklong trip across the country with adult shad in 1871, Seth Green modestly declared that "it was the longest and hardest [trip] and required more skill than any one man made before or since."[17]

Shipping fertilized eggs was easier, though still risky. The eggs were allowed to age for a couple of weeks at the hatchery where they were spawned, then placed on layers of wet moss and enclosed in a box with a block of ice. In this manner, at least some of the eggs might survive for the week or weeks that it took to send them across the continent or around the world. It was a hit or miss operation, one that often ended when the receiver opened a box to find shriveled and dead eggs, but it also succeeded more often than one might expect.

And despite all of the dangers and difficulties, Stone and his crew achieved marked success, not only at propagating the local rainbow trout, but also in distributing them. In half

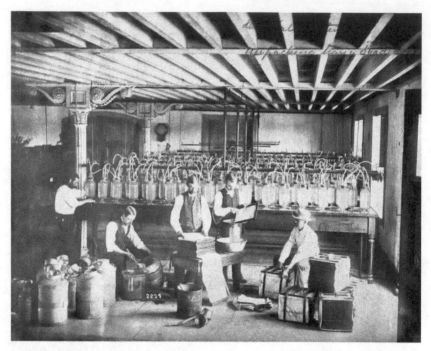

In the last decades of the nineteenth century, U.S. Fish Commission employees received and distributed fish eggs from all over the world at Central Station in Washington, D.C. The egg-hatching jars in the background are very similar to those in use today.

National Archives, College Park, RG 22 FFB B491

a dozen years, the U.S. Fish Commission had sent these fish to thirty-three of the thirty-eight states then in the Union, from Georgia to Maine and across the Midwest to Colorado and Wyoming. The Fish Commission also sent rainbow trout to England, France, Germany, and Switzerland, as well as Canada and Mexico. Many of the people who received these eggs raised them in hatcheries, established breeding programs of their own, and later shipped some of their eggs to other regions.[18]

SPENCER FULLERTON BAIRD died in the summer of 1887. His death dealt a severe blow to Stone and his work on the Pacific Coast. Baird had always treated the employees of the agency

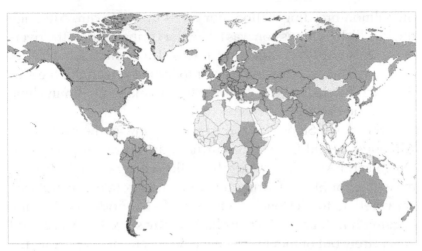

Rainbow trout, native only to the Pacific Rim from Mexico to Kamchatka, have been widely introduced and now thrive in countries all over the world (those shown in dark gray).

with respect and had long granted Stone tremendous autonomy in his West Coast operations. Stone and other members of the staff had revered Baird, in turn, almost as a father figure.

The new head of the commission, Marshall McDonald, was a former Confederate soldier and proud graduate of the Virginia Military Institute. He ruthlessly imposed his will on the staff and threatened to oust those, like Stone, who did not fit into a centralized command structure. One of McDonald's first acts was to order Stone to close down the McCloud River trout hatchery. Before becoming commissioner, McDonald had been in charge of a fish hatchery in Wytheville, Virginia, where rainbow trout from the McCloud River had been raised since 1882. McDonald declared that a sufficient breeding stock of these fish had been established in that hatchery, at another federal hatchery in Northville, Michigan, and in the breeding ponds of various state fish commissions. There was therefore no need to continue operations on the McCloud.[19]

Stone gamely obeyed the order. In June 1888, he and his crew released the breeding stock, probably about five thousand fish, into the McCloud and brought the serviceable equipment to the salmon-breeding station four miles downstream. Although the U.S. Fish Commission was no longer trying to stock the rivers of the East with Pacific salmon, it was still operating hatcheries for these fish and releasing them into local rivers under the idea that such hatcheries were the only way to maintain diminishing runs in the Pacific.[20]

Still poor, Stone was forced to take any job he could find. Although he did have some further adventures in places like Oregon and Alaska, the years that followed appear to have been miserable for Stone. During this time, he wrote nearly a dozen abject letters to McDonald, setting aside his pride and literally begging him for a job. "I have had no pay since last March and am entirely out of money and have no means of supporting my family this winter except by selling part of my real estate, which I cannot do except at a great sacrifice," read one of his letters. "I am therefore compelled to tell you of my great and pressing need of getting to work again."[21]

Finally, in 1892, after thoroughly humiliating the man, Mc-Donald allowed Stone to return to the salmon hatchery he had founded on the McCloud. Stone spent five more years there. And then, with his son ready to begin high school and his wife longing for the culture of the East where she had grown up, Stone requested a transfer. He was given a job at a federal hatchery at Cape Vincent, New York, where he and his family remained until he retired in 1906.[22]

Stone's last years were not the golden ones he might have hoped for. According to his granddaughter, Stone was forced in his retirement to rely on his son, Ned, to keep the family financially afloat. As Stone and his wife increasingly lost their mental faculties, it became an increasingly difficult job, a task that ultimately put Ned in tremendous debt. Sadly, when Stone died on Christmas Eve, 1912, he had very few of his cognitive abilities left. Somehow Ned scraped together the necessary funds, and had his father buried in the historic Mount Auburn Cemetery in Cambridge, only a few blocks from where he was born.[23]

As for the Wintu Indians of the McCloud, neither Stone's

good faith efforts nor all of the assistance they had provided to the federal government at the hatchery did them much good in the end. They received the same shabby treatment so many tribes have come to expect from the United States government. They were pushed off their land, saw agreements violated, and lost the means to pursue their traditional way of life. The approximately twenty-five hundred Wintu who remain in the area around the McCloud today are still struggling to gain federal recognition as a tribal entity.[24]

The salmon runs of the McCloud, once so thick that they sustained an entire people and were the defining feature of the ecosystem, have largely disappeared. As elsewhere, overharvest, dams, and the effects of logging, mining, and agriculture on water quality all played their part. The winter and spring Chinook salmon of the river now occupy a place on state and federal threatened and endangered species lists, and the fall run depends heavily on hatchery supplementation for its continued existence.[25]

The fish breeding stations that Stone founded so many years ago are gone now too. They lie entombed under the waters of Shasta Lake, a massive reservoir created when the Bureau of Reclamation completed the Shasta Dam on the Sacramento River above Redding, California, in 1945.

Stone's rainbows, though, are still with us. He may not have been the first to propagate and distribute rainbow trout from the wild. He wasn't the last, either—state and federal governments have been experimenting with new stocks from different regions ever since. But in most states, the first rainbows anyone ever saw or bred or released into the wild—the first rainbows anyone ever caught—were from Stone's operation on the McCloud. He popularized the rainbows. Given the way broodstocks are shared to this day, and the interbreeding that is so common in most hatcheries, there's a good chance that if you catch a rainbow today, it can trace at least a part of its ancestry back to a fish that Stone and his men captured on the McCloud those many years ago.[26]

 F I V E

A New Variety
of Trout

So just what is a rainbow trout? That depends on whom you ask, and maybe even when. A taxonomist will likely give you a Latin name for the species and, if there is a big enough cocktail napkin handy, might even draw a branch of the tree of life looking something like this:

Kingdom: Animalia
  Phylum: Chordata (animals with a dorsal nerve cord)
    Class: Actinopterygii (the ray-finned fishes)
      Order: Salmoniformes
        Family: Salmonidae (all of the present-day salmon and trout)
        Genus: *Oncorhynchus*
          Species: *Oncorhynchus mykiss*

The important point is that scientific taxonomy today is based on evolution; each group contains all of the species and only the species that had a common ancestor. The relationships have been worked out by comparing visible (but often esoteric) features on extant animals, fossils, and more recently, DNA. Throw in big chunks of guesswork and speculation to bridge the gaps, and a fairly complete tree emerges. And as you move upward, you are tracing the organism's evolutionary history back in time.

Robert Behnke has been studying, writing, one might even say obsessing, about the trout and salmon of North America since he entered graduate school in 1957. He's written books on the topic, rediscovered species that were long thought extinct, and mentored untold numbers of fisheries biologists at Colorado State University. Not everybody always agrees with his conclusions about fish taxonomy and management, but in a world of uncertainty, his word is the next best thing to receiving stone tablets on a mountaintop.[1]

According to Behnke, the story probably goes something like this: About 100 million years ago, during the late Cretaceous period, when most of the continents were beginning to take on their present-day form and dinosaurs still had another 30 million years or so to stomp the earth, the Salmonidae family diverged from the other ray-finned fishes.

About 30 or 40 million years ago, the genus *Salvelinus* (which today includes the brook trout, lake trout, and arctic char) split off to form their own group. Then, sometime during the Miocene, about 15 million years ago, at a time when most modern families, including everything from wolves to whales, had emerged, the *Salmo* genus (which today includes Atlantic salmon and brown trout) split from the *Oncorhynchus*. *Salmo* evolved in the Atlantic Ocean, *Oncorhynchus* in the Pacific.

Two to three million years ago, the fish fauna of the American West began to take on its current form. Catfishes disappeared, and sunfishes nearly so. About the same time, the cutthroat trout (*Oncorhynchus clarkii*) lineage split from the rainbow trout. The cutthroats occupied the freshwaters and ocean near the Columbia River, while the rainbows swam in waters farther south, probably around the Gulf of California. However, it was the Wisconsin glaciation, beginning about seventy thousand years ago, that distributed these fish to the places they were found before humans began rearranging them.

The Wisconsin glaciation refers to a series of three so-called ice ages that occurred between seventy thousand and ten thousand years ago in North America. During these ice ages, massive ice sheets covered parts of North America, dramatically reshaping the topography as they waxed and then waned during the interglacial periods. During this time, watersheds were constantly in flux. In some cases, ice dams blocked rivers and created large bodies of water, such as Lake Missoula. This lake, about half as large as Lake Michigan is today, burst through its dam about forty times in a two-thousand-year interval, releasing massive floods that reshaped present-day Washington as they flowed to the Columbia River gorge. In other cases, rivers shifted course or were captured by other watersheds due to changes in topography and erosion.

The shifting watersheds allowed the cutthroat trout to spread inland before barrier falls cut off many of the major tributaries of the Columbia River around fifty thousand years ago. Meanwhile, rainbow trout began to spread from south to north along the Pacific coast. They entered the Sacramento–San Joaquin river basin (which includes the Central Valley of California and the McCloud River). Sometime later, the rainbow trout began to spread up the Columbia as far as they could, pushing the cutthroats out in certain places, interbreeding with them in others. Only the cutthroats above barriers like Shoshone Falls on the Snake River and Albeni Falls on the Pend Oreille were completely safe from the invaders.

Those rainbows that first entered these river systems are known today as redband rainbows. Other rainbows, known as coastal rainbow trout, spread north around the Pacific Rim until they reached Kamchatka Peninsula. Coastal rainbows also entered the Sacramento, San Joaquin, and Columbia River systems sometime after the redbands.

The long and the short of it is this: The crazy-quilt distri-

bution of trout in the freshwaters of western North America resulted from tens of millions of years of fluctuating topography
and shifting river systems that occurred at the same time as a
series of invasions—first by the cutthroats, then by the redband
rainbows, and finally by the coastal rainbows.[2]

Making the picture even more interesting is the fact that
both rainbows and cutthroats have two different life forms: a
variety that spends all of its life in freshwater, usually close to
the place where it was spawned, and a slightly more adventur-
ous variety that, in the case of the rainbows, is known as steelhead. Although the latter spend their first one to three years of
life in freshwater streams, they eventually migrate to the Pacific
Ocean, where they roam for thousands of miles in search of sustenance. After a year or two in the ocean, they return to spawn in
the stream in which they hatched, often measuring two to three
feet long and weighing twenty pounds or more.[3]

How these two different life forms evolved is unclear and
will likely remain mysterious for some time to come, even if
molecular biologists are able to find the gene or genes that cause
some fish to head for the open ocean and others to remain at
home. However, it has led to some serious taxonomic difficulties.
Unlike the steelhead, resident rainbow trout may reach sexual
maturity when they are no more than eight inches long and
weigh less than a pound. Yet because they are the same species,
resident rainbows and steelhead occasionally interbreed, thereby
transferring genes from one population to another. And because
they commonly return to the stream in which they hatched to
spawn, steelhead are usually more closely related to the resident
rainbows in their spawning streams than they are to other steelhead in other parts of the same watershed.[4]

IF ALL OF THAT SEEMS confusing, however, consider the
plight of the early taxonomists of these fish. Many of them had
already been born before the idea of an ice age or continental ice
sheets had been publicly proposed. Most of them had grown up

and acquired much of their education years before Darwin and Wallace unveiled their theories of evolution by natural selection. Gregor Mendel's ideas and experiments on genetics did not become well known until 1900 and were not fused with Darwin's theory of evolution until the 1920s and 1930s. This means that molecular techniques such as DNA sequencing, the foundation of modern taxonomy, were beyond even science fiction. The idea that continental drift could make mountain ranges rise and fall and cause oceans to open and close was not proposed until the first part of the twentieth century and did not acquire widespread credibility until the plate tectonics theory was proposed in the 1960s.[5]

They lived, in short, in a static world, where many people still believed species were immutable and that God had placed them in their present form and location. Given such worldviews, given the fact that many regions of the world had yet to be explored and many populations of fish had yet to be discovered, and given the spectacular diversity of life forms among fish like the rainbow trout, it should come as no surprise that trout classification was, for many years, in constant flux.

A twenty-nine-year-old German naturalist named Georg Wilhelm Steller was the first to describe rainbow trout for a Western audience. Steller arrived on the Kamchatka Peninsula in 1740 as part of the Great Northern Expedition, a Russian undertaking that sought to evaluate the natural resources of a region over which that country had only recently obtained a margin of control. Steller's adventures were legion—he traveled up and down the peninsula and even partook in a brutally trying expedition to North America via the Bering Strait. His explorations included episodes of shipwreck and survival that approach those of Ernest Shackleton in their power to awe. More importantly, perhaps, he also discovered more than thirty species new to Western science and today has at least five animals named after him, including a manatee-like creature that once lived in the Bering Sea (Steller's sea cow), the heaviest eagle in the world (Steller's sea eagle), a sea lion (Steller's sea lion), a duck (Steller's eider), and a close relative of the blue jay (Steller's jay).[6]

This drawing of a rainbow trout was based on a specimen caught by Livingston Stone in the McCloud River in 1881.

H. L. Todd, 1896 USFC Bulletin, p. 230.

Steller's notes, many of them written during the winter while he was shipwrecked on a remote island in the Bering Sea, describe his discoveries with such accuracy and detail that his words are still commonly in use today. Among them was a fish that some of the natives of Kamchatka called *mykyhs*. "It is a very lively fish," he declared, "and, in my estimation, tastes better than any other fish on Kamchatka except for the king salmon."[7]

When Steller made his discovery, however, the taxonomic system in use today, including the convention of two-part Latin names for each species, was not yet standardized. It was not until 1792 that another naturalist, based on Steller's notes, gave the rainbow trout its first official Latin name: *Salmo mykiss*. The genus, *Salmo*, was the same genus in which all the salmon resided, and the term *mykiss* was based on Steller's description of the name used by the natives of Kamchatka.[8]

Because of the difficulty of establishing which fish should be included in that designation, however, as well as other factors including the limited means of communication and even the egos of the naturalists studying the problem, rainbow trout and their relatives received many other names over the next two hundred years. Members of the Lewis and Clark expedition sustained themselves on steelhead, a fish they called salmon trout, on their journey along the Columbia in 1805 and 1806. They never gave their find a Latin name, though, and certainly couldn't have been expected to recognize the fish Steller had described from halfway around the world.[9]

About fifty years after the Voyage of Discovery, a U.S. Army surgeon and naturalist named George Suckley took it upon himself to classify the North American salmonids. Suckley was a member of one of the survey crews seeking the best route for the transcontinental railroad, and during his travels he caught and preserved many different species of fish. It was Suckley who first used the term *Oncorhynchus,* meaning hooked snout, as a genus name for several of the Pacific salmon. Unfortunately for his reputation, Suckley concluded that because they looked so different, the males of various Pacific salmonids were different species from their female counterparts, and left the females as well as the rainbow and cutthroat trouts in the *Salmo* genus.[10]

The female salmon of the Pacific joined their male counterparts in the *Oncorhynchus* genus shortly after Suckley published his treatise in 1861, but confusion remained about the rainbows. Shortly after arriving in California, Livingston Stone seems to have believed the trout of the McCloud were a different species than the rainbows from farther south. "Supposing that I discover a new variety of trout, or a trout that has not been named, should I be entitled to give it a name?" he inquired of Baird in 1875. Stone dropped the matter shortly thereafter, acknowledging that the "new" trout "may not amount to anything after all." But over the next century, other taxonomists took up the challenge where Stone left off. They split the rainbows into as many as sixteen different species or combined them into a single species, depending on whether they were splitters or lumpers. And although it was questioned as early as 1914, the rainbow trout and their close relatives, the cutthroats, remained in the *Salmo,* separated from the Pacific salmon for another 127 years.[11]

AT THE JUNE 1988 MEETING OF the American Society of Ichthyologists and Herpetologists, two University of Michigan fish biologists made the case for reclassifying the Pacific salmon and trout. Evidence accumulated over the past few decades, they argued, made it impossible to justify splitting up the rainbow trout and the Pacific salmon any longer. These fishes should all be lumped together in the same genus. There were some polite and even probing questions, but they met with little true resistance from the biologists in the audience. A committee charged with standardizing the nomenclature for fishes in the United States unanimously agreed that the name should be changed and in the next edition of the standard reference work, *Common and Scientific Names of Fishes from the United States and Canada,* the trout were renamed *Oncorhynchus mykiss.*[12]

It was only after the public learned of the decision that a controversy erupted. One of the Michigan biologists who had urged the change was told he'd be shot if he visited the Pacific Northwest. The warning was likely said in jest, but nevertheless, it does reflect the tremendous anger that the genus change generated.[13]

For among that subset of anglers who were devoted to (some might even say obsessed with) catching steelhead this was not just an esoteric scientific question; there were grave issues at stake. For one thing, the steelheaders had long fought to maintain the status of their quarry as a game fish, rather than a commercial species that could be unceremoniously scooped up in nets like the other Pacific salmon. Lumping them together under the same genus made it more difficult to argue that special laws should apply to the steelhead. More importantly, though, the *Salmo* just seem to have a certain cachet. Izaak Walton, the patron saint of fly-fishing, dubbed them "the king of fresh-water fish." And in fact, anyone who has fished for a *Salmo* can trace their lineage back to the anglers of the Roman Empire.[14]

Change the genus, and all this is gone. One writer declared that the steelhead was now an "Asian fish." Another well-known fishing writer caustically and only half-facetiously noted that "it's easy to see how an organization calling itself the American Society of Ichthyologists and Herpetologists would have little compunction about changing a time-honored, smooth, and esthetically pleasing name like *Salmo* to an awkward, clunky, tongue-sticking name like *Oncorhynchus*." For the angler, he declared, salmon and steelhead would forever be held "separate and apart, no matter what their evolutionary history."[15]

In truth, in shape and appearance, even in terms of their life history, the rainbows do have more in common with *Salmo* like the Atlantic salmon and the brown trout than they do with the Pacific salmon. For one thing, the rainbows, even the steelhead, are iteroparous, like the *Salmo*, meaning they can spawn more than once. Only one or two out of every ten steelhead typically survive long enough to spawn twice, but nevertheless, this stands in stark contrast to the Pacific salmon, which all die within weeks of spawning.[16]

Restrict the analysis to the freshwater rainbows and browns and the split becomes even more counterintuitive. Surely these fish, so long known as trout, have more in common with each other than they do with the so-called salmon—Atlantic or Pacific. They are smaller, have a similar body shape, and, for the last hundred years anyway, can often be found in the same streams. But under the new classification system, the words "trout" and "salmon" no longer have any taxonomic meaning. They'll remain in use, no doubt, and well they should. There are other reasons to group things together under a common name besides their evolutionary relationship. But because, consciously or subconsciously, most people today assume that animals that share a common name are more closely related than those that do not, these terms will also undoubtedly continue to cause confusion for years to come.[17]

On top of the controversy over the genus, there was also a somewhat more civil debate about which fishes should be considered subspecies of *Oncorhynchus mykiss* and which are separate species altogether. Behnke, for example, has long considered the beautiful golden trouts of the Kern River system in California to be subspecies of the rainbow, while the American Fisheries Society considered them an entirely separate species until 2004 and keeps open the possibility of doing so again. As

Behnke himself acknowledged, such decisions are bound to be somewhat arbitrary and are likely to change as more evidence becomes available in the future.[18]

Every authority seems to agree, though, that *Oncorhynchus mykiss* does include a number of different subspecies. Because of their wide range and the fact that some populations have been isolated for thousands of years, the fish that are included under this name are quite diverse in their appearances and life histories.

But putting all of that aside, imagine a band along the Pacific Rim, from Baja California and Mazatlan all the way around to the Kamchatka Peninsula. It stretches inland along the Sacramento and San Joaquin Rivers in California, up the Klamath and into the Oregon desert, up the Columbia to the Kootenay River in northwestern Montana and to Shoshone Falls on the Snake River in Idaho, and into the Athabasca River in Alberta. That band is the native range of rainbow trout, *Oncorhynchus mykiss*.[19]

Define Me
a Gentleman

Think of domesticated species like corn, or sheep, or dogs, and you probably think of them in terms of how useful they are to humanity. Consider the other side of the equation, though, as scientists and other thinkers have begun to do. These species have benefited from the arrangement at least as much as we have. Through backbreaking labor and industrial efficiency, we have served as protectors and providers for these species, transporting them throughout the globe. Teosinte, a wild grass that begat corn, now survives in only a few nooks and crannies in Mexico and Central America, while its descendant occupies almost 5 percent of the continental United States alone. And while their wild forebears are rapidly disappearing from the earth, there are probably more domesticated sheep and more dogs born each year in the United States than humans.[1]

In the race to populate the earth, though, all three of those animals lose by a wide margin to the rainbow trout. For every one of the approximately 4 million Americans born each year, state and federal hatcheries hatch, raise, and stock approximately twenty rainbows into the country's public waters. They aren't just eggs or recently hatched fry, either; most of those fish are mature adults. And that doesn't even include the 60 million pounds of rainbow trout raised in private hatcheries every year.[2]

Perhaps even more impressive than the number of fish is the number of places they inhabit. Rainbow trout are native to an arc of territory along the Pacific Rim stretching from northern Mexico to Alaska and around to the Kamchatka Peninsula in far eastern Russia. But since the 1870s, they have been introduced to every state in the United States and to at least eighty different countries on every continent except Antarctica. Most of the introductions occurred in the last decades of the nineteenth century, and not all of them were successful. Nevertheless, rainbow trout have become a global species, both physically and culturally.[3]

Rainbow trout were the state fish of Colorado and Utah until 1994 and 1997, respectively, even though they were not native to either state. There is a movement to make rainbows an "honorary" indigenous species of South Africa, whatever that means, and they can be found on postage stamps from Malawi to New Zealand. The range expansion that corn, sheep, dogs, and humans only achieved over thousands of years, rainbow trout have accomplished in little more than a century.[4]

How did the rainbows do it? Just as with the other domesticated species, there were probably two factors. First, they were predisposed to life under the care of humans. As Robert Roosevelt noted, they could withstand warmer temperatures and more difficult conditions than other trout, and they grew more rapidly in the hatcheries. Even more importantly, though, they satisfied a powerful human need.

IT HAS BEEN SAID THAT in both quality and quantity, fishing literature surpasses that of any other sport. And if that literature is any indication, most anglers like to believe that in casting their line and watching it for a strike, they are experiencing a primal moment, a moment of expectation, a moment of life and death, a moment that is central to their being. Many writers have tried to trace the experience into prehistory, and claim that we have an innate predilection for the activity, that evolution has irrevocably installed fishing in our DNA.

Maybe these ideas have some basis in fact. It goes without saying that people have been killing and eating other animals for eons. Cro-Magnons were probably fashioning fishhooks out of wood tens of thousands of years ago, and other *Homo sapiens* used stone and bone for the same purpose. Easter Islanders may even have made hooks out of human thigh bones.[5]

And consider the prefix Anglo- or Engl-, as in Anglo-Saxon or English. These words derive from the Latin name for an ancient Germanic tribe that migrated from the Jutland peninsula, across the North Sea to Britain sometime around the fifth century A.D. Why did the Romans bestow this moniker on them? *Ancus* means hook in Latin. Some postulate that the name derives from the fact that the Angles' homeland curves into the ocean like a fishhook. But since this topography is really visible only from the air (and frankly doesn't really look like a hook at all), historian Timothy Rawson has posited another explanation: the Romans called them the Angles because they were wont to fish with a hook, instead of a spear or net as many of their contemporaries did. English speakers think and communicate, in other words, in a fishing language.[6]

And while fishing for recreation may not have as long a history as fishing for sustenance, Rawson has traced it back thousands of years. Ancients from Confucius to Augustus enjoyed fishing with hook and line, and the Egyptians appear to have been fishing for sport for at least a thousand years before Marc Antony and Cleopatra had their famous outing on the Nile. In England, a recreational fishing treatise commonly attributed to Dame Juliana Berners dates to the end of the fifteenth century, and Izaak Walton penned his famous work, *The Compleat Angler*, only about 150 years later.[7]

But in truth, the recreational fishing of today—trout fishing in particular—rests largely on a foundation built in the nineteenth century. Another historian, Foster Rhea Dulles, has written a history of recreation in America from the first colonies to the 1960s. The book describes a country that has pendulated from one end of the spectrum to the other in its attitudes toward leisure in general and fishing in particular.[8]

On one extreme were the colonists like those in the Massachusetts Bay Colony whose laws forbade anybody to "spend his time idly or unprofitably under pain of such punishment as the court shall think meete to inflict." The law specifically directed the constable to look out for "unprofitable fowlers," which presumably meant anyone out hunting birds for pleasure, and would also have applied to anyone who seemed to be enjoying the use of hook and line (let alone catch-and-release fishing). Perhaps their aversion to such enjoyments was simply a matter of survival for a people who believed their toehold on the new continent would be lost if they did not exert themselves for the greater good in every waking moment. It likely also stemmed from the fact that hunting and angling were becoming popular leisure activities among the aristocrats, Anglican clergy, and otherwise well-to-do of England—the very people and the very lifestyle the Puritans were fleeing.[9]

On the other hand, fish supplied a critical source of protein for the hard-pressed colonists, and fishing was at least better than alternatives like bear baiting, a popular pastime in England at the time, in which a bear tied to a rope was forced to battle hunting dogs, with a great loss of blood and life on both sides. Many colonists were probably never as opposed to fishing as the lawmakers in places like the Massachusetts Bay Colony might have liked. And even there, what strictures there were against fishing eventually gave way, so that by the end of the seventeenth century even Calvinist ministers could openly declare themselves to be anglers; some, like Cotton Mather, even made a point of using fishing metaphors in their sermons.[10]

By the middle of the eighteenth century, most legal restrictions and even cultural disapproval of sports and recreation had disappeared in the colonies. Along with horse races, cock-fights, and other entertainments, Philadelphians enjoyed fishing parties on the Schuylkill River. New Yorkers could fish any of a dozen or more streams full of brook trout without ever leaving Manhattan.

Then came the American Revolution, and once again, everything associated with the English aristocracy was suspect, especially exorbitant displays of idleness. In the Articles of Association, drafted by the first Continental Congress in 1774, the representatives declared that from then on, they would "encourage frugality, economy, and industry, . . . and will discountenance and discourage every species of extravagance and dissipation." Ministers well into the next century preached about the sinfulness of any sort of idleness in ways that would have made the Puritans from 150 years earlier very proud indeed. Visitors to the United States were nearly unanimous in the dismal picture they painted. "In no country that I know is there so much hard, toilsome, unremitting labor," wrote one such observer, "in none so little of the recreation and enjoyment of life."[11]

All work and no play also spelled trouble for recreational fishing. It wasn't outlawed and it didn't disappear—anglers in both the North and especially the South no doubt continued to visit their favorite waters—but they no longer flaunted it as they once had. During this era, reminisced one influential writer, "a man who went 'gunnin or fishin' lost caste among respectable people just about the same way that one did who got drunk." And when the American minister George Washington Bethune published an edition of Walton's *Compleat Angler* in 1847—a book, incidentally, that was virtually unheard of in the United States until that time—he did so anonymously. The "American Editor," as he styled himself in the book, was apparently fearful of censure if he had admitted to enjoying such an activity.[12]

Then, almost overnight, the pendulum swung back again. In the decade that preceded the Civil War and especially in the years that followed, fishing not only became acceptable again, it became a status symbol.

NEW YORK STATE HIGHWAY 27, also known as the Sunrise Highway, follows the south shore of Long Island. About half-way between Manhattan and the Hamptons, an inconspicuous brown sign on the side of the road marks the entrance to the Connetquot River State Park. Most cars zip by at sixty-five miles per hour—faster when there is less traffic, often slower when there is more. Few motorists notice the sign, and it's not hard to see why. The highway is surrounded by the same monotonous chain of tract housing and big-box stores that can be found along most of the major traffic arteries in the rest of the country, and most people have their eyes locked on the bumper in front of them anyway. There is almost nothing to give away the fact that what is today a dangerously precipitous off-ramp was once the entrance to the exclusive country retreat of many of the wealthi-est and most powerful people in the world—men with names like Tiffany, Vanderbilt, Morgan, Carnegie, and Roosevelt.

The South Side Sportsmen's Club was founded in 1866 when a group of wealthy New Yorkers purchased a tavern on a rough dirt road that was at one time the primary highway connecting Manhattan to the eastern part of Long Island and the villages along the south shore. The tavern had stood at the junction of the road and the Connetquot River since 1812, near a mill and millpond that predated the Revolution.

The tavern had long attracted the elite and powerful men from New York City and beyond. Daniel Webster and Henry Clay reputedly warmed themselves by its fire, and, by some accounts, every president of the United States had eaten fish caught in the nearby millpond. When the original owner died and his son decided to sell the property, a group of men, most of them former patrons from New York City, decided to purchase the property and convert it into a private club.

The attraction, for these men, of places like Long Island is not hard to understand. As a terminal moraine, a vast pile of de-bris deposited by the ice sheet that pushed south from present-day Canada during the last ice age, Long Island is character-ized by its sandy soil. Covered by leaves and pine needles, the ground in that area has a way of muffling footsteps and gener-

At the end of the nineteenth century, the South Side Sportsmen's Club counted among its members many of the wealthiest and most powerful men in the world. Most lived in New York City and visited the club on weekends to hunt and fish. They began raising McCloud River rainbow trout in their private hatchery in 1880.

Courtesy of the Dowling College Archives and Special Collections

ating a peaceful quiet. Natural springs abound, feeding streams that flow gently over a sandy bottom, and oaks and pines whisper in the dependable ocean breeze.

Visit it today, and the scene is much the same. Gladed oaks and pines frame a view of the venerable old wood-shingled building that was once a tavern and later served as the clubhouse for the South Side Sportsmen's Club. The millpond still lies just below the large front porch. Inside, behind a locked door, is the bar where members of the club would retreat after a day of fishing to drink brandy, warm themselves by the stove, and boast about the one that got away. The smell of taxidermy and old wood in the closed-up room is intoxicating. Dotting the walls are black-and-white pictures of tweed-clad men holding cigars and various types of fish and game.

It wasn't just the scenery or the brandy, though, that drew these men to the Connetquot. The popular sporting magazine

Forest and Stream declared that Long Island was "the greatest trout country in the State," and that the Connetquot was "one of the finest trout streams on Long Island." In 1865, the Connetquot had been the first river in America ever stocked with brown trout, which are native to Europe. And the members of the South Side Sportsmen's Club made sure it was also one of the first to receive Stone's rainbows, stocking it with fish it received from the U.S. Fish Commission in 1880.[13]

In a cabinet in the clubhouse, the members kept fish registers, where every day they recorded their catch. The pages have become yellowed and fragile now, but they are still there. Even allowing for some exaggeration, there is no question that the fishing was excellent. In 1896, a fairly typical year, the one hundred members of the club caught sixteen thousand trout, including brookies, browns, and rainbows.

And even though it is now a public park, the scenery remains much as it did a hundred years ago, and the fishing is still unbelievable. The hatchery established by the club is still in operation, churning out hundreds of thousands of rainbow, brook, and brown trout every year. For a fee, anybody can stand along the quiet sandy banks of that beautiful spring-fed stream and catch dozens of trout in only a few hours. "It's unlike anywhere else in the world," the owner of a nearby fly shop said when I visited one summer afternoon. "It's catching, not fishing."

WHY WAS IT THAT fishing so quickly garnered such a following after the Civil War, especially among wealthy urban Americans? Perhaps it was just an excuse to get away from the city. New York and other urban areas in the last decades of the nineteenth century were filthy and unhealthy places. Factories, slaughterhouses, stables, and tanneries spewed pollutants into the water and the air. Garbage and dead animals often clogged the sewers, emanating an awful stench, and water supplies were frequently contaminated with human waste. And while the conditions were often the worst in the crowded slums, the more well-to-do could hardly avoid the effects altogether. Diseases like smallpox, measles, typhoid fever, malaria, diphtheria, and tuberculosis often broke out into epidemics that carried away rich and poor alike. About one of every four babies died before they reached their twelfth month in New York in 1868, and that was hardly an exceptional year.[14]

Perhaps fishing became so alluring simply because people do not appreciate what they have until it is lost. The nineteenth century saw an explosion in America's urban population. Whereas less than 7 percent of Americans had lived in cities at the opening of the nineteenth century, 33 percent lived there by the end, representing 30 million people. Rampant industrialization and urbanization, of course, doomed fishing in most of these places. Robert Roosevelt complained in 1865 that "streams in the neighborhood of New York [City] that formerly were alive with trout are now totally deserted. The Bronx, famous alike for its historical associations and its once excellent fishing, does not now seem to hold a solitary trout, or indeed fish of any kind."[15]

And it wasn't just the growth of the cities that gave Americans a sense of loss. Until 1891, the Census Bureau had officially designated a frontier line, beyond which the country was considered unsettled. But when the superintendent of the census for 1890 examined the maps, he decided that the scattered settlements in the American West made it impossible to draw this line any longer. Jumping off from this conclusion, the historian Frederick Jackson Turner gave a speech in conjunction with the Chicago World's Fair in 1893 in which he declared that "the frontier has gone, and with its going has closed the first period of American history." Americans would never again have the opportunity to test and revitalize themselves in the wilderness, he opined, and would therefore have to find other ways to promote individualism and democracy. Turner's speech, perhaps the most influential (and controversial) ever given by an American historian, electrified the nation.[16]

On the other hand, perhaps recreational fishing exploded

onto the national scene at the end of the nineteenth century because, all of a sudden, previously remote and unfished waters were easily accessible to those who could afford a train ticket. Although the first rails had been laid down only fifty years earlier, a web with more than one hundred thousand miles of track covered the United States by the 1880s. Fishing magazines were filled with stories by men who had taken the last train out of town on Friday evening for a weekend of fishing in some unexplored stream.

One writer determined that "it has been ascertained to an almost mathematical nicety that it will cost the metropolitan angler one dollar for every pound of trout he takes, no matter where or under what circumstances he fishes." It would cost less to travel to streams close to the cities, but they were likely to be fished out. A more expensive ticket to more distant places would give access to waters where the fish "swarm in such abundance as absolutely to embarrass the angler." Those who could afford it usually chose the latter.[17]

Indeed, the tickets sold to anglers became such a lucrative source of income for the railroads that many went to great lengths to advertise the quality of the fishing along their lines. One railroad announced that it would give twenty dollars in gold to anyone who caught a rainbow trout weighing ten pounds or more along its route. And a couple of anglers a year actually did so. Almost all the railroads offered free transportation to the state and federal fish commissions, on the understanding that some of the fish being transported would be stocked into the streams along the route. Some railroads even built hatcheries so they could keep the waters they serviced well stocked.[18]

Many of the wealthiest anglers bought up previously remote land to turn into private clubs like the South Side Club, and for those who could not afford to join a private club there were dozens of newly minted public resorts. And while the angler could imagine himself a fearless explorer of untamed lands during the day, he did not have to do so at night. One Chicago and Northwestern Railroad guidebook, for example, boasted that "the resorts in the North Woods have been kept as close to the wilds as possible without sacrificing a single material comfort that might add to the pleasure of the traveler and sportsman." As one wit has declared, "Contrary to Frederick Jackson Turner's famous argument, the frontier was never closed; it was just reopened as a tourist attraction."[19]

Perhaps fishing regained prestige because it offered a means for the newly rich to establish their credentials. Although the names may evoke old-money respectability today, Rockefeller, Carnegie, Morgan, Vanderbilt, and others earned their fortunes in the decades that surrounded the turn of the twentieth century. Initially scorned by the established upper crust of New York, these nouveaux riches went to great lengths to prove their gentility and gain admittance into high society. And fishing was by no means the least important means of doing so. Angling had long been associated with the aristocracy in England, lending it a certain cachet among a subset of Americans who had not fallen under the strictures of the Puritan resurgence. Moreover, to be an "angler" meant not just that you had time for leisure and money to purchase expensive equipment, train tickets, and sometimes even membership in a private fishing club. It meant you had the knowledge and refinement necessary to understand the natural history and aesthetics of a beautiful stream; it meant you had the skill needed to cast a nearly weightless fly onto a glassy pool without the trout ever noticing the horsehair tippet attached.[20]

Perhaps fishing achieved prominence because, like George Perkins Marsh a decade earlier, the well-to-do men during the industrial revolution feared for their virility. In the years that followed the Civil War, blacks from the South joined an influx of immigrants from southern and eastern Europe to swell the northern cities. The physical vigor of these newcomers, who did much of the manual labor, disquieted the Americans of northern European descent who had earlier established themselves as America's ruling class. More familiar with the classics than we are today, they feared they were following in the footsteps of the Roman Empire, or perhaps the French aristocracy. They feared that success and wealth generated an effete ruling class that would easily and inevitably be overrun by the uncivilized hordes. Images of the Goths and the guillotine plagued their thoughts.[21]

In addition to an unprecedented level of racism, such fears prompted a backlash against the "overcivilization" of the "Nordic" race. Oliver Wendell Holmes, Sr., for example, warned that "such a set of black-coated, stiff-jointed, soft-muscled, paste-complexioned youth as we can boast in our Atlantic cities never before sprang from the loins of Anglo-Saxon lineage." Teddy Roosevelt cautioned that northern Europeans were committing "race suicide." And one New York doctor named George Miller Beard diagnosed a new disease among the well-to-do characterized by a "lack of nerve force." He called this depressed malaise "neurasthenia," and the term quickly entered the popular lexicon.[22]

Several solutions were proposed. Some urged Anglo-Saxon women to have more children. Beard believed it was simply a matter of zapping the sufferers with electricity to restore the energy reserves of their central nervous systems. But while both of these solutions may have had their adherents (indeed, many of the World War I veterans admitted to psychiatric hospitals were subjected to Beard's electrotherapy), others believed the race could be revitalized only if it reconnected with its uncivilized roots.[23]

The outdoor life became de rigueur. Tourists inundated camps in the Adirondacks and other wilderness resorts. Hardier souls scorned the resorts and went camping in the most remote regions they could find. But to really reconnect with their virility, men needed to capture and kill. Some took up hunting, which Teddy Roosevelt called a suitable activity "for a vigorous and masterful people," a sport that required "energy, resolution, manliness, self-reliance, and a capacity for hardy self help." The majority of those who took up field sports, though, looked to the "finny tribe," as they often liked to describe their quarry. One

New York Times article in 1874 concluded that "perhaps no description of outdoor sport is more generally followed or admired
than that of trout fishing." It was a sport that reinvigorated men
with a healthy touch of the primitive, and yet at the same time
allowed them to retain their identities as gentlemen.[24]

Perhaps the burgeoning popularity of fishing was a logical outcome of movements and philosophies that had emerged
on both sides of the Atlantic over the previous century. There
were the Primitivists, who believed civilization had alienated
people from the wisdom, morality, and invigorating influence
of the natural world. There were the Romantics and the Tran-
scendentalists and the Deists, who were convinced that nature
offered the best place to see and connect with God. Together,
the thinkers behind these movements gave the out of doors a
powerful, mystical, and often subconscious aura that persists to
this day.[25]

Others have noted that the entire concept of childhood
changed toward the end of the nineteenth century. The notion
that children were simply imperfect adults was flipped on its
head. As one writer declared in 1896, "scientists have informed
us that the child alone possesses in their fullness the distinctive
features of humanity, that the highest human types as represented in men of genius present a striking approximation to the
child type." The iconic image of the barefoot boy with a straw
hat on his head and a fishing pole over his shoulder was a cre-
ation of the last decades of the nineteenth century—a perfect
expression of the idea that children were closer to the primitive
and thus closer to nature, God, and happiness. It was an idea
with staying power, an idea that is with us to the present day. As
Sigmund Freud wrote in 1929, "The feeling of happiness derived
from the satisfaction of a wild instinctual impulse untamed by
the ego is incomparably more intense than that derived from
sating an instinct that has been tamed."[26]

Perhaps the most intriguing explanation along these lines
comes from historian T. J. Jackson Lears. When he was a Ph.D.
student at Yale University in the 1970s, Lears set out to exam-

ine American culture in fin-de-siècle America. Textbooks at the time, he recalled, told of "an optimistic, energetic society about to reach the full vigor of industrial maturity." But after poring through piles of books, magazines, and private letters from the late Victorian era, what he found were "loud lamentations among the people who called themselves 'the leadership class'—complaints that elites were rotting from within at the precise moment they were threatened from without by working class unrest. . . . [I]t involved a kind of cultural asphyxiation among the educated and affluent, a sense that bourgeois existence had become stifling and unreal."[27]

Lears set out his findings in a now classic book, *No Place of Grace*. In addition to many of the societal changes described above, Lears proposed that Americans experienced a seismic shift in their worldviews in the nineteenth century. As religion and community became less important, personal happiness became a goal in itself—a new development in Western Christianity. Ironically, though, the more people pursued this psychic harmony, the more elusive it became. Culture became "weightless," as Friedrich Nietzsche put it. As a result, people began searching for a new source of meaning through intense experience, rather than through the self-denial of the Protestants. Their search led many people to the out of doors, to the strenuous life preached by men like Theodore Roosevelt. And while Lears did not dwell on fishing, it is not hard to extend his argument. As untold numbers of angling writers have noted, fishing not only allows people to test themselves against the natural world, it literally offers a direct and quivering line to the ultimate intense experience, the struggle for life or death.[28]

OF COURSE THE MEN WHO escaped the cities for a weekend of angling at the South Side Sportsmen's Club and in other public and private waters were not looking to catch just any old fish. Through innumerable letters to the sporting magazines in the last decades of the nineteenth century, these men hashed out a rough hierarchy for the fish of the world. The most basic division was between the "game" fish and the "rough" or "coarse" fish—distinctions that are still used to this day.

"Define me a gentleman and I will define you a 'game' fish," wrote the editor of *Forest and Stream* in one of his fishing guidebooks, "which the same is known by the company he keeps, and recognized by his dress and address, features, habits, intelligence, haunts, food, and manner of eating. The true game-fish, of which the trout and salmon are frequently the types, inhabit the fairest regions of nature's beautiful domain. They drink only from the purest fountains, and subsist upon the choicest food their pellucid streams supply. . . . [It] is self-evident that no fish which inhabit foul or sluggish waters can be 'game-fish.' It is impossible from the very circumstances of their surroundings and associations. They may flash with tinsel and tawdry attire; they may strike with the brute force of a blacksmith, or exhibit the dexterity of a prize-fighter, but their low breeding and vulgar quality cannot be mistaken. Their haunts, their very food and manner of eating, betray their grossness."[29]

Perhaps the lowest of the low were the catfish. Not only were they found in sluggish waters where they scavenged in the mud, but they were also a favored food for the slaves of the old South. Gentlemen would never intentionally try to catch such a fish, and if they did so by accident, some anglers refused even to touch them, preferring to cut the line and kick the undesirable back into the river.[30]

The bass, on the other hand, had some staunch and prolific defenders. Some saw these fish as quintessential Americans. "The national fish of America is the black bass," wrote one of their champions. "It is self-reliant, and when placed in new waters not merely makes itself at home, but appropriates the locality, explores its furthest recesses and devours its aboriginal inhabitants." Even bass fishermen, though, usually acknowledged that salmon and trout were the noblest of fish. "The trout is an aristocrat of his race, a gaudy gallant, a swell fellow in fish circles, yet game and clever as your highbred gentleman should be," admitted one angler. "The bass may be taken to represent the sturdier upper class."[31]

Among the aristocrats, there were further distinctions. The Atlantic salmon had the advantage of being the royal fish of England, but having disappeared from most American waters by the end of the nineteenth century, they were not often caught or discussed. Brown trout from Germany and Scotland, which were being introduced to the eastern United States about the same time as the rainbows, also held a place near the top. But the two fish that received the most ink in the sporting periodicals were the brook trout and the rainbow. The brook trout were often associated with innocence, boyhood, and the early years of the Republic. The brook trout, wrote one angler, is "a creature of cold brooks and little singing tributary streams. He loves the gentle ripples, deep, dark haunts beneath the roots of overhanging trees, the catacombs of drift piles, the erosion chambers of the banks and the shade of overhanging bushes. The rainbow is quite another trout. He is the aristocratic buccaneer of big waters. His heritage is the deep, boiling rifts, swift currents, and dancing whirlpools. Angling for rainbows in the vortical tumult of racing water is trout fishing raised to its supreme eminence."[32]

Indeed, within a few decades, rainbows were often acknowledged as the best fighters of the whole family. "The rainbow trout is without a doubt one of the finest game fishes we have. It fights desperately until completely exhausted and leaps again and again," wrote Theodore Gordon, one of the most influential American angling writers.[33]

And because the character of the quarry was considered a direct reflection of the character of the pursuer, rainbows quickly became a favorite among sportsmen. "The man who has ever hooked a leaping fighting rainbow on light tackle in a canoe on the Soo Rapids will have all the thrills that should come to the honest fisherman," opined another angler, "and if he wins, his manly chest is the proper place for the pinning of all the medals that are the reward of victory."[34]

FISHING DIDN'T SIMPLY reflect the race and class conflicts of the time, either. Sometimes it caused them. It was the captains of industry who were perhaps most responsible for the degraded and polluted state of the urban centers, and it was they who were often, ironically, the only ones who could afford to escape. Certainly they were the only ones allowed to join exclusive fishing clubs like the South Side.

As many of the best streams passed into the private hands of the urban elite, rural residents from Long Island and elsewhere found themselves excluded from waters they had long fished and perhaps relied upon for income and sustenance. Even in public waters, many sportfishermen regarded fishing for food — potfishing — to be a vulgar and ignorant practice. They justified their actions as a means of protecting wildlife that would otherwise have disappeared from overfishing and overhunting. They proudly compared their lands to royal tracts like the New Forest in England. And with their power and influence, it was easy for them to get laws passed that enshrined their own behaviors and made other practices much more difficult.[35]

Their actions, though, generated a tremendous backlash. Some people expressed their resentment in letters to newspapers and journals. "These gentlemen know they can't have in this country an 'aristocracy of blood,' so they want one of the very meanest kind — one of money," ranted one disgusted angler in 1876. By locking up the land and the fish, they would "make the poor man hate the rich man," he opined, and "make a very large majority of men who take any interest in hunting or fishing, feel that they will do all they can to destroy the fish and game entirely, rather than a few men who may chance to have a little money should have all of those privileges."[36]

And indeed, many locals simply ignored the laws and the No Trespassing signs, ultimately prompting the members of the South Side Sportsmen's Club and other keepers of private waters to hire armed guards. Violent encounters between the poachers and the "watchers" were not uncommon. Men were shot and beaten to within an inch of their lives. Barns burned, boats sank, dogs were killed. Long Island in particular became known for its fearless and wanton poachers.[37]

To this day, scholars disagree about whether nineteenth-century sportsmen were the saviors of America's fish and wildlife or just a powerful special interest group appropriating the resource for their own use. I don't know which side is right. I'm biased toward public access and empathize with the locals who were pushed out of their favorite streams. But on the other hand, thanks to the members of the South Side Sportsmen's Club, the area around the Connetquot was spared the concrete and asphalt that swallowed up much of the rest of the region. Today even plebeians like me can go fishing there. And certainly the fish, especially the rainbow trout, benefitted.[38]

 S E V E N

Paying Customers and Hatchery Product

I doubt it's as satisfying for the fish, but whatever they're missing by being artificially spawned is at least partially offset by the evident satisfaction that people like John Riger derive from doing it for them. When I met him, Riger was managing the Crystal River Hatchery, an unprepossessing Colorado Division of Wildlife facility about thirty miles down valley from the glitter of Aspen, Colorado. On the clear but bitterly cold November day when visited, Riger stood waist deep in a concrete raceway with two assistants, spawning rainbow trout. The fish were reasonably big—they were at least eighteen inches long and probably weighed a couple of pounds—and they lazily circled around in a galvanized steel cattle tank that contained a mild anesthetic.

In a routine that they repeated over and over, the three men reached into the tank, secured a female, and lifted her out. Sliding thumb and forefinger firmly along her belly, they stripped the luminescent orange eggs into a black plastic pan about the size and shape of a dog's water bowl. They then reached in for a male—how they determined the sex from such an angle remains a mystery to me—and repeated the process. This time, though, it was white milt that sprayed into the bowl, fertilizing the eggs almost instantly.

Although the process was methodical, it was not, for Riger, just a routine. With what I think was the utmost sincerity, Riger repeatedly held up fish for me to observe and asked whether it wasn't the most beautiful thing I had ever seen. Apparently I responded with a sufficient amount of awe, because after a while, I was allowed to put on some waders, climb into the raceway, and try spawning a few fish myself.

When we were done, about a quarter of a million eggs floated in a large bucket. We carried them to a shed where they would rest for several weeks until they became eyed eggs, so called because a black spot appears on the embryos. At that time, the eggs would be run through the egg picker, a sort of Ferris wheel that picks up one egg at a time and examines the color with an electronic eye. Eggs that have not been successfully fertilized or have, for some reason, died, typically turn white. A sideways puff from an air gun blows the white eggs, known as blanks, into the reject pile while another gun blows the keepers into a different tray.

Except for those saved for future breeding, no fish are hatched or raised at the facility. The state of Colorado has special rearing units for that purpose. And so, shortly after their trip through the Ferris wheel, the eggs are packed in a cooler, placed in the back of a truck, and taken to a rearing facility that they will call home until they are big enough to stock.

The Crystal River Hatchery spawns about 10 million trout every year, more than any other hatchery in the state. And so, if you catch a trout in Colorado, there's a good chance that it began its life there, that it was brought into being, and even in some ways designed, by John Riger.

IN A 1939 REPORT, the head of fish culture for the federal government declared with some pride that technology now made it possible to produce "an entirely 'synthetic' fish." Not only could fish culturists artificially spawn and raise these animals, but they could manipulate their genes through breeding programs and even by hybridizing different species. As Robert Roosevelt had predicted as far back as 1872, "What was done with the common tomatoes, potatoes, onions, and hundreds of other vegetable productions . . . may in a higher degree be carried into effect with fish."[1]

And so it was. Rainbow trout seem particularly amenable to such artificial selection programs, adapting rapidly over just a few generations. Over the decades, rainbows have been bred to grow faster, mature earlier, and breed at different times of the year. Culturists have tried to select for disease resistance, fecundity, and even such things as color, shape, and fighting ability.[2]

Because these fish have been shipped back and forth from hatchery to hatchery in an attempt to introduce new genes and produce new breeds, it has become impossible to keep track of their lineage. Which is not to say that people haven't tried. One intrepid group of government fish squeezers (as the fish culturists sometimes refer to themselves) has put together a document known as the National Fish Strain Registry. In it, more than seventy-five strains of rainbow trout are listed and ranked according to obvious traits like when they spawn, their growth rate, and their tolerance for things like handling, crowding, temperature, and disease.[3]

Want a rainbow that spawns in May? Try the Colorado River strain. Need one that can tolerate high temperatures? The London strain is for you. Do your fish have a problem with infectious hematopoietic necrosis? Might want to breed them with some Reynoldsdales—they're highly resistant to that disease.

The strains are also ranked according to behavioral traits, such as their tendency to migrate, a particularly vexing issue for everyone who has ever tried to stock rainbows. It appears that beginning with Livingston Stone, culturists commonly bred

resident rainbows with peripatetic steelhead, so that the resulting fish had a tendency to wander far and wide. It was not a
trait that endeared them to anyone stocking a private section of
stream, like the members of the South Side Sportsmen's Club, or
even to state fish and game officers who were never sure what
happened to their fish. The problem has largely been solved
today, for their wandering habits have been bred out of them
through generations in hatcheries, but culturists still keep their
eye on it.[4]

Perhaps most intriguingly, the trout strains also get ranked
according to their "angling susceptibility." Rainbows have
long been popular with fisheries managers because they are
the Goldilocks of trout in terms of their catchability. Perhaps
because they have evolved during the hundreds of years that
people have been dangling hooks in front of them, brown trout
are notoriously difficult to catch by that method. Brookies and
cutthroats occupy the other end of the spectrum, relatively un-
discerning and easy to catch. Rainbows, though, occupy the
middle ground. Easier to catch than the browns, they are not
fished out as quickly after stocking as brooks or cutthroats.[5]

Often, managers take advantage of the different traits to
create a sort of fish cocktail. In Crowley Reservoir, in California,
the Department of Fish and Game stocks three types of rainbows: Kamloops, which tend to be easily caught from the shore,
but seldom live long enough to achieve any great size; Colemans,
which prefer deeper water and are often caught by trollers but
don't overwinter very well; and Eagle Lake Rainbows, which
are generalists and often survive for a sufficient number of years
to become trophy-sized fish.[6]

Some fish and game departments like to play public relations games with their fish. Some zealous managers try to get the
attention of anglers and the press by stocking albino rainbows,
or better yet, Centennial Golden rainbows, an eye-popping
orange-colored variety of rainbow trout first bred in the hatcheries of West Virginia. Others go for size. Every year, California
stocks a very few whoppers: rainbows weighing, rumor has it,

up to ten pounds. The likelihood of catching one is not great; you probably have a better chance of winning the lottery. But every so often somebody does, and, one hotel operator in the eastern Sierras told me, "before noon the phone is ringing from people in southern California making reservations."[7]

My fish, as I tend to think of the trout I helped to spawn, were Bellaire rainbows. They aren't anything exotic, but their report card is enough to make any parent proud. On a scale of 1 to 5, where 1 is "poor" and 5 is "superior," they get 5s for their ability to handle stress and tolerate crowds. They typically weigh about a third of a pound at one year old. That's bigger than almost all of the brooks, browns, and cutthroats, as well as most of the other rainbow strains. They get a 4 in "Angling Susceptibility," and an incomplete in "Tendency to Migrate." Apparently no one knows.

WHEN THE STATES BEGAN creating their own fish and game commissions after the Civil War, they typically funded them through the state treasury. Each year, the commissioners had to request another disbursement. In an era of small government, budgets were typically tight, and the amount allotted to these new agencies was minimal. In 1880, therefore, the editor of the popular sporting journal *Forest and Stream* floated a new idea. Speaking for many of his readers, he declared that hunters would be willing to pay a small fee for the right to hunt if the revenue was used to pay for game wardens.[8]

The proposal didn't catch on right away, but over the next few decades, hunting and fishing licenses gradually became the standard means for funding fish and game agencies. The licensing system satisfied the powerful urban sportsmen because it gave them a sense of ownership over the fish and wildlife, animals that often persisted only in rural areas far from their doors. It satisfied the politicians because it removed the burden of funding these agencies from the state's annual budget. And for the rainbow trout that were by that time swimming in state

hatcheries from New York to California, it was this idea, perhaps more than anything else, that ultimately enabled them to
conquer the freshwater world.[9]

Why? Consider this proposition: Governments like those
in the United States typically enact laws based on the premise
that the administering agencies will act as rational, objective servants of the public good. In reality, though, agencies are staffed
by human beings. And consciously or unconsciously, humans
often make decisions on the basis of their own self-interest. This
may seem cynical or self-evident, depending on your point of
view. But either way, it is one of the central premises behind a
field of economics known as public choice theory, a field that
earned several of its founders the Nobel Prize. It is a proposition, in other words, that has some real bona fides.[10]

And there is little question where self-interest lies if an
agency's budget and the employee's paychecks depend on the
number of people who go fishing. Limits and regulations might
preserve the fishery enough to maintain license sales at a certain
level, but it's no way to expand the revenue stream. Better, as one
nineteenth-century U.S. fish commissioner put it, "to expend a
small amount of public money in making fish so abundant that
they can be caught without restriction, . . . than to expend a
much larger amount in preventing people from catching the few
that still remain after generations of impoverishment."[11]

More fish, more anglers, more licenses. There was a compelling connection that was further reinforced in 1950 when
Congress passed the Dingell-Johnson Act. With this law, the federal government began collecting a 10 percent tax on all fishing
equipment sold in the United States. Once collected, these funds
passed back to the state fish and game agencies for their fishery
programs.[12]

So pervasive and unquestioned was the paradigm that
agency higher-ups unabashedly compared themselves to the
producers of such things as "beef cattle" and "toffee." As recently
as 2001, a Colorado wildlife commissioner told a credulous reporter that the Colorado Division of Wildlife "is a business, and

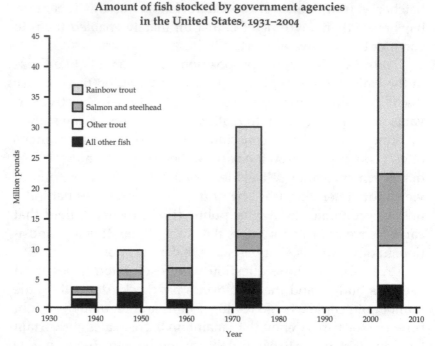

**Amount of fish stocked by government agencies
in the United States, 1931–2004**

State and federal agencies increased both the number and size of the fish they
produced over the course of the twentieth century. By 2004, these agencies were
stocking more than 40 million pounds of fish in the freshwaters of the United
States, close to half of which were rainbow trout.

we need to run it like that." Sportsmen were the "paying customers" and fish were now called a "hatchery product."[13]

This is not to say that the fish and game agencies did not
serve other functions as well. They did. Agencies implemented
and enforced rules about the number of animals that could be
taken, the seasons in which they could be pursued, and the
means by which they could be caught or killed. But if the amount
of money they spent is any indication, hatcheries were far and
away their most important function.

In its first dozen years, the U.S. Fish Commission spent
more than 75 percent of its ever expanding budget on fish
propagation. As state agencies grew, they too devoted much

of their revenue to building and operating hatcheries so that, ultimately, their expenditures for this purpose dwarfed those of the federal government. In 1948, for example, the states raked in more than $19 million from the sale of fishing licenses. They spent more than 85 percent of that income building and operating more than five hundred hatcheries and only about 5 percent on fisheries research. Similarly, the federal government spent about $2.5 million operating ninety-nine hatcheries and only about $400,000 on fisheries research.[14]

It wasn't just agency officials who liked to build hatcheries, either. Thanks in large part to heavy doses of propaganda from the fish and game agencies, many people came to see hatcheries as an easy technical fix for depleted fisheries. Anglers came to expect their favorite waters to be stocked on a regular basis. And as local economies grew to depend on those hopeful anglers, entire communities began to count on monthly infusions of fish. Agencies have done little to discourage this line of reasoning. As recently as 2005, the U.S. Fish and Wildlife Service justified its own operations by publishing an economic analysis of its stocking program. Among other things, the document declared that every dollar spent growing and stocking rainbow trout resulted in thirty-two dollars of economic activity through everything from worm sales to airplane fees.[15]

There was some pushback, right from the beginning; a few economists and biologists sought to discredit the economic justifications of the U.S. Fish and Wildlife Service and other agencies, but they didn't get very far. It had become a sort of feedback loop. The more the agencies stocked, the more demand there was for stocked fish, the more the agencies spent on growing fish. And it wasn't just the stocking programs that drove the hatchery boom. The hatcheries themselves were easily transformed into pork, a good way for politicians to show their constituents they could bring state or federal dollars to their district. Perhaps Bernard Shanks, a former director of the Washington Department of Fish and Wildlife, put it best. After his proposal to close several hatcheries was met with howls of protest, Shanks declared that a fish hatchery is what you get "if you cross a sacred cow with a military base."[16]

BECAUSE MANY OF THE rivers in California experience prolonged dry spells interspersed with occasional floods, they have long posed a challenge to that state's fish and game department. As an example, just before the outbreak of World War II, the head of the department told colleagues from other states what he declared was a "more or less true story."

"Take, for example, a sunny Sunday morning in May. Mr. Los Angeles looks out of his window and for no good reason at all discovers that there has been a cloudburst on the desert the night before and there is water in the Los Angeles River. By 9:30 o'clock, 20,000 telephone calls have come to the Fish and Game Commission to come out and plant some fish because there is water in the Los Angeles River. Since we have one of the most efficient departments in the country, by 10 o'clock a truckload has started out. We carry a siren on the trucks, by which, at the end of the planting, we let everybody know that the planting has been accomplished. By 11 o'clock the fish are caught out of the stream, and at noon the river has dried up again!"[17]

Which parts of the story fall into the less true category and which into the more is not entirely clear. What is true is that by the late 1930s, California was stocking much of its hatchery product as "catchables," fish that were large enough to satisfy an angler who caught them immediately after they had been stocked. Typically, this meant something in the seven-to-ten-inch range, occasionally a little longer. And California wasn't alone. Many other states were also employing this strategy, especially in waters that were heavily fished.[18]

It wasn't always that way. In the previous century, the primary goal of the state and federal fish commissions was to introduce desirable species to waters they had not previously inhabited. This meant transplanting trout from larger lowland

California rivers into high Sierra lakes that had hitherto been devoid of any fish at all. It meant shipping fish from the West to be planted in the East and, more commonly, shipping fish from the East to be planted in the West.[19]

The idea was to scatter the fish into as many waters as possible and hope that at least a few survived and reproduced to create a new fishery. "Should the Commission make a success of a single river of a size, or half the size of the Sacramento," Livingston Stone once declared of the Pacific salmon he had shipped east, "it would pay for all that has been expended in the direction on all the other waters of the United States."[20]

Fish eggs and recently hatched fish known as fry, therefore, were the most commonly stocked. Not only were they much cheaper to produce, they were also easier to transport. They could be poured into an old tin can, slipped into a mule's saddlebags, and hauled up to high alpine lakes. Even more importantly, tens of thousands of them could be carried in the specially designed train cars that the U.S. Fish Commission and many state fish commissions owned. A few of the eggs and fry would be parceled out at each stop to anybody who wanted to try stocking a new species in their local stream or lake.[21]

As time went by, however, the goals shifted, while the techniques remained largely unchanged. More and more, people saw stocking as a means of increasing the total number of fish in a particular water, instead of a means of seeding a new self-sustaining population. Few questioned whether the fry they were stocking were contributing anything to the fishery; research on the matter was close to nonexistent. Once the fish left the hatchery they were out of sight and, as the saying goes, out of mind.[22]

Finally, in the 1920s, fifty years after the U.S. Fish Commission first began stocking fish, some of the state agencies began tracking the fish that they had stocked. To their consternation, they found that in almost every case, their plants of eggs and fry had done nothing to increase the number of fish caught by anglers.[23]

And so, a new metric was born. The percentage of the stocked fish that were ultimately caught by anglers became known as "return to creel." Instead of measuring success based on the number of fish they produced and at what cost, fish and game agencies began assessing return to creel, and the cost of each fish that was caught. During the Great Depression, employees of relief programs like the Civilian Conservation Corps were put to work surveying anglers about their catch. And when the results showed that larger fish had a much higher return to creel, they were put to work building the much larger hatcheries that were needed to raise fingerling and, ultimately, catchable fish.[24]

But while the concept of return to creel may have improved the economic efficiency of stocking programs, it also had some unintended consequences that persist to this day. Managers quickly discovered that one way to improve their score on this measure was to advertise where the fish were being stocked. California had sirens on its stocking trucks by 1939, and in places like Colorado to this day, the latest stocking locations are only a phone call away. Managers also concluded that return to creel could be improved by dumping all the trout in an easily accessible roadside location, instead of using boats to spread them throughout a stream.[25]

Best of all, both in terms of return to creel and the justification for intensive stocking, were the millions of ponds and reservoirs that were created on everything from the tiniest creeks to the mightiest rivers in the middle decades of the twentieth century. Whatever their nominal purpose, they often created fantastic trout habitat both above and below the dam—often, in the West at least, turning warm, muddy desert rivers into something that could fool a trout into thinking it was in a cold, clear mountain stream or lake. The reservoirs frequently had limited spawning habitat for natural reproduction and weekend sportsmen could easily access them. They became magnets, in other words, for fish-stocking trucks.[26]

Some agencies even built their own reservoirs for the sole

purpose of attracting anglers. Take the U.S. Forest Service. In the early part of the twentieth century, the agency had primarily viewed the national forests in utilitarian terms. They were sources of timber, grazing land, water, and minerals. But in 1916, Congress passed the National Park Service Organic Act to "conserve" certain lands from such uses and to provide for their "enjoyment" by the people. The increasingly mobile American public quickly became enamored with the idea of such natural "pleasuring-grounds," and members of Congress allocated increasing amounts of money and land to the National Park Service. Alarmed by the competition, Forest Service administrators quickly jumped on board. Agency officials who previously would have scoffed at the idea soon found themselves boasting about the recreational possibilities on their lands as well, trying to poach some of the Park Service's newfound constituency. Not only did they encourage state fish and game agencies to stock the waters in the national forests, occasionally they even built dams for the sole purpose of improving trout habitat.[27]

And so that quintessential image of fishermen lined up shoulder to shoulder, yanking trout out of the water almost as fast as they could be poured from the truck, was born. To be fair, at least some agency personnel disliked such scenes. But since most fish, even catchables, die within a few weeks of being stocked anyway, it was hard to defy the logic of the system. Those who objected had to hold their noses and go along with the program. Stocking catchables worked, as one fish and game commissioner from the state of Connecticut put it in 1938, if you "keep it under control and face the facts by deliberately treating it as a manufacturing proposition."[28]

IN THE YEARS THAT FOLLOWED World War II, America experienced a period of economic growth and prosperity unlike any that had occurred before. Wealth spread to a broad spectrum of Americans who, at the same time, were enjoying an unprecedented amount of leisure time. The twelve-hour day, typical of the early nineteenth century, had largely given way and eight-hour workdays had become standard. Thanks largely to the efforts of the labor unions, many Americans were even enjoying two-day weekends and paid vacations for the first time. And thanks to the technology and manufacturing abilities developed during the war, Americans could cheaply purchase the easiest and most efficient fishing gear ever developed. Spinning reels, fiberglass rods, and long strands of monofilament made it easy even for novices to catch fish. Americans responded by making fishing one of the most popular sports in the country. One person in five, about 21 million people, went fishing in 1955, together spending 400 million days trying to catch fish, a huge increase over the prewar years.[29]

Almost as important, the end of World War II also yielded an abundance of surplus military airplanes as well as a large number of demobilized pilots. Forty-year-old Al Reese was the first to join the California Department of Fish and Game. A former barnstormer and crop duster, Reese spent the war years training army cadets to fly. When that gig was over, Reese turned his can-do mind to another problem, stocking California's abundant and often remote mountain lakes. He was sure he could do it from the air.[30]

First, Reese tried freezing the fish in ice blocks and parachuting them in ice cream containers. Both of these techniques, though, proved dangerous and difficult. And so, one day, Reese and his assistants tried a simpler technique. They put fifty trout and some water into a five-gallon can and threw it out the window toward a hatchery pond about 350 feet below. They missed, and the can bounced along the rocks nearby instead. But when observers recovered the twisted metal debris, they found sixteen fish still swimming in the small amount of water that remained. It was a stunning result for fishery managers who had long been telling anglers not to throw fish back, but to gently place them back in the water.[31]

Assured by this mishap that the fish could survive the impact, Reese set out to discover whether they could make the trip

Al Reese and Carrol Faist (third from left) began California's aerial fish-stocking program with the World War II surplus Beechcraft C-45 pictured here. This photograph was taken in 1950 or 1951.

Courtesy of Carrol Faist, California Department of Fish and Game

without the protection of tin and water. Reese and his partner grabbed some more fish, hopped in a vehicle, and hit the gas until they were moving seventy-five miles per hour down the hatchery road. At that point, the men grabbed fish, one by one, and held them out the window for two minutes, at which point they pulled them back in and dropped them back into the water. And once again, the fish survived.[32]

These and other experiments were enough to convince Reese and his superiors that his plan could work. Reese persuaded the department to purchase a military surplus C-45 transport plane and also hired another pilot named Carrol Faist, a man who had flown forty missions on B-24 Liberator bombers in the Pacific. One July day in 1949, Reese and Faist set off with a plane full of trout for their first drop into an actual alpine lake. While one of them flew the plane (history does not record who had which job), the other went into the back, loaded up a hopper full of fish, and peered through a four-inch-by-four-inch hole cut in the bottom of the plane. As soon as the lake was visible through the hole, the bombardier released the fish. The sudden reduction in weight caused the plane to bounce twenty feet higher, making it a tricky and dangerous job for the man in back. Nevertheless, the drop was a success.

Observers on the ground described a cloud of mist that suddenly appeared behind the plane, full of the barely distinguishable dark shapes of small fish. After hanging still for a moment, or so it seemed, the fish tumbled through the air in a spray of water and splashed like raindrops in the middle of the lake. Many of them, according to the observers, survived.[33]

That's not to say it was a pleasant experience for the fish. Dropping out of a plane that was about two hundred feet in the air and traveling at a speed of around two hundred miles per hour (typical of planting runs both then and today), the fingerlings would have hit the water with a vertical speed of about thirty miles per hour. Decades after Reese and Faist first dropped their fish, I talked to a biologist who witnessed a similar event while snorkeling in one of the lakes of the Sierra Nevada. Many of the fish were ripped in half on impact, he told me, and many others were so stunned they immediately sank to the bottom, never to recover.[34]

Nevertheless, Reese and Faist considered their experiment a success, and they became even more confident with each run. Except when the bombardier missed (yes, it's happened) and the fish landed in the trees, they found that fish dropped from planes actually survived better than fish that had to make the trip bouncing around for hours in a can on the back of a mule. It was cheaper too. It only cost about four dollars to stock one thousand fish from an airplane, compared to about twenty dollars by other methods.[35]

For the record, California wasn't the first to drop fish from the air. Other innovators in Quebec and New York had experimented and even employed the technique on a small scale before and during the war. But within a year, California had eclipsed their efforts and shown the world what the future of fish stocking would look like. They even tried other animals as well.[36]

They dropped beaver outfitted with special parachutes, as well as turkeys and partridges. On occasion they dropped shrimp and aquatic plants into the lakes they had just stocked to provide food and cover for the trout. But fish were by far the most common species to make the trip. By the end of the 1950s, California and many other states were routinely using airplanes and helicopters to stock the backcountry. Thousands of previously fishless lakes were soon full of trout. It was a boon, no doubt, for the millions of recreational anglers who emerged after World War II. But even more, it was a boon for the trout.[37]

AS FOR THE RAINBOWS I had helped spawn in November, I visited them again about eight months later at their new home, not too far from Aspen, Colorado, in a facility known as the Chalk Cliffs Rearing Unit. They'd changed quite a bit from the last time I saw them, but still, they were no bigger than my pinky. They'd moved into a rundown trailer, a throwaway from the Department of Corrections, but it had been retrofitted with everything a young fish could possibly desire. They had a long tank, with a constant flow of fresh water, plenty of food, and no predators.

Nevertheless, according to Chris Hertrich, the forty-year-old ex-Marine who managed the unit, survival for these fish was a flip of the coin. Unlike some hatcheries, which get their water from springs, at Chalk Cliffs the water comes from a nearby creek. That means temperature fluctuations that can stop the growth of the fish when it gets cold and kill them when it gets too warm. It means diseases and parasites can never be eliminated, and often necessitate treating the entire system with toxic chemicals like formalin. And when I was visiting, Hertrich was struggling to contain the damage from a mudslide several miles away that had turned the water supply dark brown and was threatening to fill the facility with sediment.

Despite such travails, Chalk Cliffs produces some seven hundred thousand catchable trout each year. The biggest fish live in half a dozen ponds that are all connected by an underground pipe, several thousand feet long, to a holding tank at the lower end of the facility. Pulling the plug has the same effect as flushing the toilet; the fish get the ride of a lifetime to the holding tank, where a retrofitted forklift scoops them up and dumps them into a hatchery truck for their journey to a stream or reservoir on Colorado's Front Range or eastern plains.

Almost a year later, a cloudy May day found me standing around the gravelly edge of a municipal reservoir near Colorado Springs, getting ready to see my fish one more time. I'd received an email from Hertrich the day before, informing me that some of them were scheduled to be stocked. Waiting with me were a group of military veterans. Most were eighty years old or more; they'd fought in World War II and Korea. It was the first time they had been out of the VA hospital for many days, perhaps weeks or months. Most seemed to be enjoying themselves, though the action was slow at best. Only a couple of fish had been caught since I arrived. And then, the truck arrived.

Within seconds of hitting the water, the fish went to work. Bobbers began going under, the folks who had volunteered to help the veterans began running around yelling "fish on" and helping to reel (as did I), and many of the fish that had been in the hatchery truck a few minutes earlier were transferred to a stringer, destined for a fish fry at the Fitzsimmons Army Medical Center that night.

It may seem strange, after all of the effort I put into it, but I must admit that I've never derived a great deal of satisfaction from catching hatchery rainbows. Indeed, it was fish like these that drove me away from the sport for many years. I grew disgusted or bored after a day of catching nothing but stockers, and for more than a decade never picked up a rod. Under ordinary circumstances, I certainly wouldn't have gained any pleasure from standing on the bathtub rings of a little reservoir, reeling in fish like these. It's hard not to enjoy being part of an expedition like that, though.

I took a couple of fish home with me, of course. One I ate. It was fine, not that flavorful, but it provided a nice base for the butter and lemon. The other I had mounted, and some day I intend to hang it above my desk. The taxidermy didn't leave much of the original fish, I'm afraid. The fins remain, the skull's still there, and I suppose there's some real skin under all the paint, but other than that it's wholly man-made. Which is, of course, entirely appropriate.

 E I G H T

A Full-Scale
Military Operation

Above Flaming Gorge, the Green River drains most of southwestern Wyoming and a small chunk of northeastern Utah. It's a big watershed. In fact, if it weren't for some fancy political maneuvering by the Colorado congressional delegation in 1921, most people would probably recognize it today as the true origin of the Colorado River.[1]

For millions of years, the Green flowed unchecked through this high alpine desert, through a country of low buttes dotted with sagebrush, through a country of seemingly unceasing wind. And when it reached the southern end, in what is now Utah, the river carved through the brilliant vermilion rocks of the Uinta Mountains, creating a classic western canyon known as Flaming Gorge. The birthplace of the Green, though, was in the snows of Wyoming's Wind River Range. And when the spring came, and the sun got higher in the sky, the locked-up water was released in a torrent, carrying huge volumes of sediment downstream. The river was so thick with mud that witnesses insisted you

94

could walk across it. Banks were scoured away and deposited elsewhere and large sandbars appeared and disappeared from one season to the next.[2]

The temperature, the scouring floods, the muddy water — together they made the Green a challenging environment in which to live, and a unique fauna developed in response. Four large fish dominated the main stem of the river. On the smaller end of the scale, though still up to three feet long, there were the humpback chub, the bonytail, and the razorback sucker, all of which have a large hump or keel behind their heads, a feature that may help them maintain position in the swift waters of the Colorado. And then there was the Colorado pikeminnow, the largest minnow in North America, a fish that could reach six feet in length and weigh up to a hundred pounds. These leviathans were the top predators in the river, eating everything from fish to unsuspecting waterfowl, and they were known to migrate hundreds of miles in search of food and spawning sites.[3]

The early settlers in the region seem to have taken some pleasure in catching and eating the native fish, especially the pikeminnow. The meat was tasty — some compared it to salmon — and catching one could be the experience of a lifetime. Anglers used rabbits, birds, whatever they could find, as bait. In an interview many years later, one old-timer recalled his grandfather tying one end of a clothesline to the bumper of a pickup, the other end to an enormous hook and a piece of chicken. When the line went taut, he backed up the truck and hauled out a pikeminnow that he claimed was every bit as big as a junior high schooler.[4]

Pickup trucks and clotheslines, though, didn't really qualify as sport, and the fish themselves weren't pretty, at least to anglers who had been trained to see trout and salmon as the standard by which all others should be judged. The natives didn't fight as long, they were bonier, and sometimes they even ate trout eggs. Suckers and chubs were especially scorned. These were rough fish, trash fish, undesirables.

In one of the first attempts to reengineer the river's fishery, in the nineteenth century, the U.S. Fish Commission introduced German carp. These fish, too, thrive in warm and muddy water. Spencer Fullerton Baird, the first commissioner, was an especially strong advocate for these hardy fish, believing they would provide a valuable food source to the growing nation. Within a couple of decades, though, the carp had fallen into disgrace. Americans never seemed to develop a taste for the fish, and farmers and anglers alike complained that they uprooted vegetation, muddied the water, provided lousy sport, and drove out more valuable species of fish. By the middle of the twentieth century, carp in the Green River had become just another one of the undesirables.[5]

And so, when the Utah and Wyoming fisheries managers learned that the Bureau of Reclamation was going to build some dams on the Green River—one at Flaming Gorge and another, known as the Fontenelle Dam, about 150 miles upstream—they smelled an opportunity. The projects, they knew, would dampen the floods, remove the sediment, and cool the water by releasing it from the bottom of the reservoirs. The river, in other words, would soon be ideal for trout.

ROTENONE IS A POWERFUL piscicide, a chemical that kills fish and other animals with gills. Our understanding of just how it functions has evolved over the years—it seems to somehow interfere with respiration—but whatever the mechanism, it works. Drop enough of it into a lake or river and, soon thereafter, dead or dying fish and aquatic invertebrates will pop to the surface. It's even effective, according to that 1954 movie classic, against the Creature from the Black Lagoon.

Various plants in the pea family produce rotenone, and South American Indians have been using their ground-up roots for hundreds of years to catch fish. American fisheries biologists began experimenting with it early in the twentieth century. Most of these applications, though, were in small ponds or lakes. Few dreamed it could be used in moving waters, where it would be rapidly diluted and washed downstream.[6]

For the generation that entered the workforce after World War II, though, few things seemed impossible. Many fisheries managers were graduates of the fish and wildlife departments that had only recently opened their doors at various universities, and they were eager to test out their new knowledge. And thanks to the Dingell-Johnson Act of 1950, which taxed recreational fishing products throughout the country and passed the funds back to the states to improve sportfishing, there was plenty of money available.[7]

First it was California, where, beginning in 1952, two ambitious managers poisoned almost three hundred miles of the Russian River and its tributaries so that they could introduce more desirable species. Montana followed suit in the Marias River and the Clark Fork. In the south, the Little Tennessee was poisoned and then filled with nonnative rainbow trout. Not even the national parks were spared. The fauna in Abrams Creek in Great Smoky Mountains National Park was erased, and it too was replaced with nonnative rainbows. All in all, in the decade that followed the Russian River experiment, managers poisoned countless waters on behalf of the rainbows and other game fish. Under the Dingell-Johnson program alone, twenty-five hundred miles of streams and 225,000 acres of lakes were poisoned in thirty-four states, and that was likely just the tip of the iceberg. Many other lakes and streams were treated using other funding sources.[8]

In 1962, therefore, the fish and game managers of Utah and Wyoming were inspired to try a massive rotenone project of their own. They decided to "rehabilitate" the Green River above Flaming Gorge by killing all of the fish in the watershed, native and nonnative alike, and then introducing millions of rainbow trout. It was a big watershed, about 15,000 square miles, about the same size as Connecticut and Massachusetts combined. There were almost 450 miles of the river and its tributaries that would need to be treated. But given what other managers had already done elsewhere, it wasn't too much of a stretch, logistically or philosophically.[9]

Again, the money part was easy. Agency personnel projected that more than a hundred thousand anglers would fish the area, spending upward of a million dollars every year. They argued that the project would be a boon to local economies and generate much more in taxes than the approximately $150,000 it would cost. And since sympathetic congressmen were easily persuaded to appropriate most of the funds for the project, it would cost the states close to nothing.[10]

To disperse the rotenone, the agencies decided to create poisoning stations along every ten miles of the Green and its tributaries. At each station, a fifty-five-gallon drum of rotenone would drain its contents into drip lines stretched across the river. The planners concluded it would be best to stagger the times at which the spigots were opened. Each station would begin operation as soon as the rotenone from the upstream station passed by, thus ensuring a consistent front of rotenone flowing downstream, killing and pushing its victims in front of it.[11]

To make sure no fish survived in the backwaters, the agencies decided to employ airboats. Once the project began, the crews of these boats would speed up and down the river, spraying extra rotenone into all the nooks and crannies they could find. A helicopter, too, was brought in, to spray any sloughs that might be disconnected from the river itself during low water levels.[12]

The Bureau of Reclamation initially planned to finish the dam and begin filling the reservoir sometime in September 1962, and the fish and game agencies planned to begin the rotenone project the moment they did so. That way, they reasoned, the poison would not flow any farther downstream—it would simply collect in the filling reservoir until time and sunlight broke it down into harmless byproducts. Once it was gone, they would introduce the rainbows.

It seemed like a sound plan until the bureau postponed completion of the dam until November. At that time of year, the water in the Green River would be cold, and, for whatever reason, cold water limits the effectiveness of rotenone; it simply doesn't kill the fish the way it's supposed to. So the fisheries managers came up with a new plan. Instead of using the dam to detoxify the rotenone, they would use another chemical, a neutralizing agent known as potassium permanganate. Nobody had ever done anything like it before, but it seemed feasible. The only question was where to set up the detoxification station.

Ultimately, they decided to use a bridge in a place known as Browns Park, just over the border in Colorado. The Coloradans didn't mind. The head of the Colorado Department of Fish and Game was fully supportive of the project, wishing only that they would allow the rotenone to run a little farther downstream and kill a few more of the undesirables in his state before they neutralized it. With a couple of exceptions, nobody seemed to give much thought to the fact that the detoxification site was only sixteen miles above Dinosaur National Monument.

BORN IN 1916 TO PARENTS who loved traveling and the out of doors, Robert Rush Miller had seen much of the West by the time he finally settled down at a boarding school in California at the age of ten. He spent much of his teenage years hiking and exploring the deserts of eastern California and western Nevada. On these sojourns through the Mojave Desert, he became fascinated by the fishes he found in nearly impossible places—like the tiny springs in the middle of Death Valley that were separated from the nearest running water by miles of scorching sand.

Miller was the type of person who was more comfortable in the out of doors and with other species than he was with other people. Hard of hearing from a very young age, Miller was socially awkward even when he was at his most upbeat. And when depression struck, as it did periodically throughout his life, he became even more withdrawn. It was a condition that eventually led him to seek help at the Mayo Clinic.[13]

Miller, though, was no shrinking violet. His fascination with the native fishes of the West gave him a singular focus that quickly drove him through a Ph.D. program and into a position of prominence in the field of ichthyology and a tenured job at the University of Michigan. At a time when scientific groups like the Ecological Society of America had firm rules against taking any position on policy matters, Miller was unafraid to speak out against the rampant dam building and other projects that threatened native species.

Not surprisingly then, when Miller learned of the agencies' plans for the Green River, he quickly took up the fight. He sought to defend his beloved native fishes the way a mother bear might defend her cubs. And just as he had in so many of his endeavors, Miller soon joined forces with another ichthyologist named Carl Hubbs.

I wish I could have observed these two men interacting, because it is surely one of the most complex relationships imaginable. Consider the ties: Hubbs had been both authority and mentor to Miller as his Ph.D. adviser at the University of Michigan. Shortly thereafter, when Miller married his daughter Frances, Hubbs became Miller's father-in-law. In contrast to the introverted Miller, Hubbs possessed a supreme self-confidence. He was a type A personality, a steamroller of a man who had become one of the foremost scientists of the day.[14]

Somehow, though, they achieved a partnership of equals that was one of the most productive in the history of the field. (A great deal of credit probably goes to their wives, Laura Hubbs and Frances Hubbs Miller, who played an integral role in all of their work.) Instead of the turf battles that typify so many academic relationships (let alone in-law relationships), Miller and Hubbs shared their ideas and their passion for the native fishes of the American West. On their countless collecting trips (almost always accompanied by Laura and Frances and frequently by their children and grandchildren), they gathered thousands of specimens, many of them species never before known to science. They generated and frequently coauthored a large body of work, all of which is still cited today.

And when they decided to try to stop the fish and game agencies from poisoning the Green River, they did so with the same zeal and thoroughness that characterized their other efforts. Neither man is alive today, but some of their offspring are, and so are some of their colleagues. Together with their letters, many of which are archived at the University of Michigan, they tell the story, a tale both of resolve and of frustration. For, although Miller and Hubbs persuaded their fellow ichthyologists in academia to pass a resolution opposing the project, and although they sent copies of the resolution along with their personal pleas to more than four hundred magazine editors, congressmen, government officials, conservation organizations, and other influential people, they got almost no traction. Nobody seemed to care about the callous destruction of the West's native fish.[15]

Miller blamed agency public relations specialists who he believed had successfully molded public opinion to view such operations as the logical, inevitable manifestation of progress. "The public is often brainwashed by vicious and wholly one-sided propaganda," Miller ranted in one letter. "Words such as 'rehabilitation' and phrases such as 'recontamination of the stream' (by *native* fishes) are used with reckless abandon."[16]

It was not just the public. With a few notable exceptions, many scientists who privately opposed the project publicly kept their mouths shut. Many were apparently fearful of government reprisals, and admitted as much in their letters to Miller. Government agencies threatened to cut funding for those universities that allowed their faculty to speak out against the Green River rehabilitation project and, in at least one case, they seem to have followed through with it. Conservation organizations were likewise curiously quiet on the issue, a phenomenon for which I have no adequate explanation, except to speculate that the movement and the thinking behind it had simply not grown large enough at that time to encompass mostly unknown fish from an underwater realm that was out of sight and therefore out of mind.[17]

As for the state and federal agency personnel who were planning the project, they largely dismissed Hubbs and Miller's concern for native fish as unfounded or out of touch. Agency officials argued that no species would be eliminated from the earth, because they would survive in other sections of the Colorado River. Miller conceded the point, but noted that the Green River contained the last sizable refuge for many of these fish and that "the noose is getting awfully tight."[18]

Around and around they went. Would any of the fish downstream be harmed, especially in Dinosaur National Monument? Was the humpback chub, found largely upstream from Flaming Gorge, a distinct species? Or was it the same species as another chub commonly found downstream? In some cases, it reached ridiculousness; at one point the combatants began arguing about whether rotenone could really be called a poison if it actually suffocated fish by constricting the capillaries in their gills.[19]

Ultimately, the debate went nowhere, because the two sides were arguing about science and logistics while their real differences were in the realm of value and worldview. Miller et al. divided the world into native and nonnative, whereas agency officials thought in the categories of sport and trash fish. Miller and his colleagues saw degradation where their opponents saw progress. And the ichthyologists expected unpredictability, while the agencies anticipated a logical and straightforward outcome to their attempt to manipulate the ecosystem.

And so, positions on both sides hardened and exasperation set in. "Dr. Miller does not understand that the public demands the maintenance of usable sport fish populations and that public officials will either work to this end or be removed from office," fumed one U.S. Fish and Wildlife Service official. "His sense of values do not line up with reality. The people of the Southwest are more than tired of well-meaning individuals promoting the preservation of everything in sight for posterity."[20]

In the end, he was probably right, at least in terms of the reality of the day. Almost everybody who knew about the operation supported it, especially the locals. Miller, Hubbs, and their few allies lost. The operation would proceed as planned.[21]

AT 8:00 A.M. ON SEPTEMBER 4, 1962, the men in charge
of the uppermost drip stations opened the spigots. Rotenone
sprayed into the river, turning it a milky white. At about the
same time, the airboats began buzzing up and down the river
while the helicopter chopped the air overhead. The project had
all the features, according to one witness, "of a full-scale military
operation."[22]

As the rotenone front progressed downstream, it drove
large schools of desperate fish in front of it, a sight that deeply
impressed itself on those who saw them "thrashing about and
struggling for air on the surface of the river."[23]

The poisoning continued for three days and nights, ultimately using more than twenty thousand gallons of rotenone. A
multitude of curious onlookers from nearby towns and ranches
thronged the banks. Assured by agency personnel that the rotenone posed no risk to human health (there is some controversy
today about whether it can cause Parkinson's disease), many
scooped up the dead and dying fish and brought them home
for dinner.[24]

By the end of the three days, about 450 tons of fish were
dead and the aquatic invertebrates were almost completely destroyed. The Green River and its tributaries were virtually devoid of visible life, an ecological clean slate. The operation, in
other words, was a complete success. Or at least it seemed that
way to most of the exhausted personnel as they packed up their
equipment and headed home.[25]

Thirty miles downstream from Flaming Gorge, though, at
the detoxification station on the Browns Park bridge, there was
trouble. Traveling only about one mile an hour, the rotenone
front arrived at the station a full day after the poisoning part
of the operation had ended. When it did, everything began to
go wrong. The potassium permanganate was the wrong consis-
tency, making it difficult to spread evenly. A cold front arrived,
causing temperatures to plummet and fierce winds to pick up.
And worst of all, for reasons that were unclear at the time, the

For three days in 1962, helicopters, airboats, and drip stations like this one
poured tens of thousands of gallons of fish poison into the Green River during
its "rehabilitation." The poison, known as rotenone, can be seen turning the river
white.

W. H. Kittams (National Park Service)

rotenone concentration in the river as it reached the bridge was
much higher than expected. There was not enough potassium
permanganate to neutralize all of it.[26]

And so, while some of the personnel roared off in their
trucks to scour the countryside for more potassium permanganate, others began thinking of ways to preserve what they had.
They decided to cut the amount they were putting into the river
to the bare minimum necessary to neutralize the chemical. But
since the most reliable method they had for determining rotenone concentrations was a bioassay—a fancy term for putting
fish in a cage, dropping them in the water, and watching to see
if they died or showed signs of distress—this was a dicey operation at best.[27]

Imagine somebody below the bridge, in the middle of the
night, with temperatures well below freezing. Imagine them

holding a flashlight, staring into wind-whipped water that had turned dark red from the potassium permanganate, trying to gauge the health of a fish submerged in a cage. If the fish looked sick, they would radio up to the bridge and tell them to throw more potassium permanganate into the river. Imagine maintaining this effort for three and a half days, and you probably have some idea of what it was like.

Needless to say, some of the rotenone got through and continued flowing downstream. And a few days later, dead and dying fish were found in Dinosaur National Monument.[28]

DEFENDERS OF THE ROTENONE project on the Green River later pointed out that the dead fish in the monument may have died upstream from Browns Park and been washed down. And later surveys showed that the abundance of fish in the monument had not been seriously reduced. Miller disputed these assertions, but either way, it almost instantly became a moot point.[29]

Less than a decade earlier, Dinosaur National Monument had become hallowed ground for many Americans, thanks to the efforts of people like David Brower of the Sierra Club. The short version of the story: the Bureau of Reclamation sought to build a dam in the monument. Brower and others believed that such a move would not only destroy a beautiful canyon, it would set a terrible precedent. National parks would be protected only until an economically valuable purpose was found for their resources. The conservationists mounted a massive and brilliant public relations campaign that convinced many Americans that Dinosaur was a sacred place. Even though most had never seen it, people from all regions of the country wrote letters to Congress opposing the project. Ultimately, Brower and his allies won. Legislators refused to approve the funding for the dam. And thanks to their efforts, Dinosaur National Monument had become a widely recognized national treasure.[30]

But there was also another factor, perhaps equally important. About three weeks after the poisoning of the Green River was completed, Rachel Carson published her epochal book, *Silent Spring*. A jeremiad against pesticides and by extension piscicides, the book quickly became a must read. Man-made chemicals joined national parks as perhaps the two most important issues in the budding environmental movement. Mixing the two in the middle of Dinosaur National Monument yielded a volatile combination.[31]

Officials with the National Park Service blasted their counterparts in the U.S. Fish and Wildlife Service for facilitating the project. Fish and Wildlife officials returned fire, issuing memos that dripped with sarcasm and ridicule, especially when Miller's name came up. Testy exchanges occurred over whether the humpback chub was a legitimate species and, if so, how many had been killed and how many remained.[32]

For his part, Robert Miller seems to have sensed an opportunity in the crisis over the national monument, an opportunity to proselytize on behalf of native fish, wherever they were found. In an incendiary article in *National Parks* magazine, Miller denounced the U.S. Fish and Wildlife Service and the state agencies for their incompetence, their efforts to mislead the public with terms like "rehabilitation," and their blind devotion to fish like the rainbow trout.[33]

Outraged letters from the public, even from Rachel Carson herself, began pouring into the offices of the U.S. Fish and Wildlife Service, the White House, and the halls of Congress. The chairman of the House Natural Resources and Power Subcommittee pointedly demanded to know why the Fish and Wildlife Service had supported the project and whether the Department of Interior truly believed it had "fully complied with its responsibilities."[34]

Finally, the national press took notice, with articles about the controversy appearing in papers like the *New York Times*. It may have been too late for the native fish of the Green River, but Miller's efforts did, in the end, pay off. The fate of native fish had at last entered the national spotlight.[35]

ULTIMATELY, RESPONSIBILITY FOR the Green River
rotenone project must land on Stewart Udall, the secretary of the
interior at the time. Udall's fiefdom included the two primary
disputants, the U.S. Fish and Wildlife Service and the National
Park Service. Udall is an interesting character. Galvanized by
Carson's book, Udall helped found the modern environmental
movement during his tenure as secretary from 1961 to 1969. He
wrote books and articles about what he called "the quiet crisis,"
and successfully pressed lawmakers to pass the Wilderness Act
in 1964 and an early version of the Endangered Species Act in
1966. But he also presided over some of the biggest dam projects
in the nation's history, and oversaw many of its extractive industries. So it's interesting to speculate: What role did Udall play in
the Green River rotenone project? And what role did the project
play in his thinking?[36]

When I spoke to this still vibrant man forty-five years later,
he did not recall the Green River episode. He noted that he had
relied heavily on his subordinates, as any secretary of such a big
department must, and that he likely did so in this case too. Other
evidence, however, suggests not only that Udall played a larger
role than he could recall, but that the poisoning was in some
ways formative for him as secretary of the interior.

When the incident occurred, Udall was a relative newcomer
to a department full of career administrators. After serving as a
congressman for six years, he was just getting his bearings in
the executive branch, and it appears that whatever misgivings
he may have had, he felt powerless or at least hesitant to use his
power to stop the project. In a letter preserved at the University of Michigan, author and national parks advocate Wallace
Stegner wrote to Robert Miller that he had spoken with Udall
and pleaded with him to block the poisoning earlier in 1962.
"Apparently the peculiar division of powers between states and
federal government did not permit any such action," Stegner
wrote. But, he added, "I know the Secretary's convictions ran
with yours, not with those of the Utah Fish and Game Commission."[37]

Shortly after the poisoning, though, Udall reviewed the matter, and whatever feelings of powerlessness he once had seem to have dissipated. In the spring of 1963, Udall issued a memorandum in which he declared that none of the agencies in his department would be allowed to fund the use of rotenone or any other piscicides again until further research had been conducted. Perhaps more importantly, though, in the same memo Udall declared that from then on, "Whenever there is question of danger to a unique species, the potential loss to the pool of genes of living material is of such significance that this must be a dominant consideration in evaluating the advisability of the total project."[38]

These were heady words, one of the first times anyone in a position of such power had articulated such a clear policy on endangered species. Indeed, shortly after writing the review of the Green River project, Udall created a new body, the Committee on Rare and Endangered Wildlife Species, for the lofty purpose of "halting the further disappearance of endangered wildlife." It is no small irony that Daniel Janzen, the man who had overseen the Green River operation for the federal government, became the head of the new committee. Nevertheless, Janzen seems to have taken the job seriously; the work of this group, and the list of species it generated, would one day lead directly to the Endangered Species Act of 1973, the law that protects these species today. And despite their recent and sometimes heated conflict over the Green River, Janzen put Robert Miller and a few of his fellow ichthyologists in charge of recommending fish for the list. The result: by 1966 the same federal agency that had been helping to eradicate native fish in the Green River only a few years earlier had become their chief protector.[39]

WHEN I BEGAN RESEARCHING the Green River incident, one of the first things I did was to browse through archived copies of the *Denver Post* that were published around the time of the poisoning. Without really thinking about it, I assumed

After losing the battle to stop the Green River rotenone project of 1962, Robert Miller (left), Jerry Smith (right), and others collected specimens of Colorado pikeminnow and other fish that were killed.

Courtesy of G. R. Smith

that a newspaper from the largest city within five hundred miles of the Green River, the largest city in the Rocky Mountain West, would prominently cover an operation that wiped out all the native fish in such a huge watershed, a watershed that included parts of Colorado. But on September 3, 1962, the day before the operation began: nothing. There were a few articles about Cuba—the Cuban missile crisis would erupt one month later—but there was no news that seemed terribly pressing. The next few days, still nothing. Finally, in the September 7 edition, on page 23, I found a twelve-sentence article by the outdoor editor. The first sentence declared: "Fishing was never better on the Upper Green River than it was early this week when Wyoming Game and Fish Department crews opened a mass fish poisoning project." He was referring to the scores of dead fish that anybody could pick up and bring home for dinner. There were a few more details of the operation and the "rough" fish that were being eradicated to make way for trout. And that was it.[40]

Working my way backward, I found another article about the operation from September 2 that likewise read like a press release from the Wyoming Department of Game and Fish. There was one sentence declaring that "several comparatively rare fish species exist in the river, and some national conservationists had objected to the poisoning on the grounds the species might be completely destroyed." But that was toward the end of the article and seemed to be offered largely as a sort of apologetic explanation for why the agencies intended to detoxify the poison before it could flow through Colorado.[41]

It wasn't just the *Denver Post*, either. The *Salt Lake Tribune* buried a five-sentence article on the eleventh page of the B section on September 2, applauding the pending elimination of the Green's "trash" fish. Wyoming's *Rock Springs Daily Rocket* celebrated the demise of the "coarse" fish, and the magazines put out by the Utah and Wyoming game and fish departments were similar.[42]

Initially I was frustrated, but it eventually dawned on me that the void said more about the incident and the times than any article ever could. It's hard to believe in this day and age, but at the time, the fact that the government planned to poison all of the fish in one of the biggest watersheds in the West, to risk extinction for many of the native species, so that it could stock those waters with nonnative rainbow trout, was simply not newsworthy. It wasn't worth the ink. Most of the editors seem to have taken their cue from a "suggested" editorial put out by the Wyoming Game and Fish Commission. "Do people go fishing for carp and suckers in Wyoming?" the release rhetorically asked. "You bet they don't. They like a creel of fat, colorful, fighting rainbow trout."[43]

What a difference a couple of decades makes. Imagine a government agency conducting such a mass-poisoning project today. Four of the "trash" species the fisheries managers tried so hard to eradicate in the Green River in 1962—the Colorado pikeminnow, the humpback chub, the razorback sucker, and the bonytail—now reside on the Endangered Species list, unlikely to be removed any time soon. Americans have now spent more than $100 million trying to pull these fish back from the brink, and the bill likely will get much higher before all is said and done.[44]

To be fair, there are those, like Robert Wiley, who defend the operation to this day. Now retired, Wiley was a rookie Wyoming fisheries biologist in 1962; he cut his teeth on the Green River project. When he describes the project in papers and speeches, Wiley emphasizes the nonnative nuisance species like the carp. In his characteristically combative manner, he acknowledges that "public opinion might require that we do something else," but insists that if the two dams were being built today, rotenoning the river would still be "biologically sound."[45]

Wiley paints a picture of "farsighted" managers who understood as early as 1959 that the dams would irrevocably change the habitat in the Green River and devastate native species accustomed to a warm, muddy, and occasionally torrential river. Given that the natives would be destroyed anyway, according to this logic, it only made sense to use rotenone to get rid of the undesirable nonnatives that might take over the reservoirs.[46]

On the other hand, there are people like Professor Gerald Smith at the University of Michigan. Smith was a graduate student with Robert Miller in 1962, and he helped monitor and collect specimens from the Green River during the rotenone project. Some time after interviewing him over the phone, I received an email from Smith. In it was a carefully written statement. Obviously not satisfied with the haphazard nature of our talk, Smith had sat down and, forty-five years after the poisoning occurred, summed up his thoughts. I wish I could reprint the entire document here, with its carefully laid-out list of the erroneous assumptions and logical, biological, and ethical mistakes made by the planners of the poisoning. Instead, I will just quote his conclusion: "The primary good that came out of the operation was to provide such a clear bad example of management, that mass poisonings have been largely abandoned as a fishery management tool," he declared. "The project's many failures helped undermine the 1950s assumptions that complex ecosystems are bad and monocultures are good, that large-scale poisonings are modern and diversity is unnecessary and old-fashioned, and that it is possible and desirable to attempt to completely subjugate Nature."[47]

I wasn't present for the poisoning. But I've read scores of archived letters and memos from the various agencies involved. These documents all seem to confirm Smith's conclusions about "1950s assumptions" more than Wiley's picture of "farsighted" managers. "Biologists regard aquatic habitats as pastures in which the end products are fishes which can be taken by man by sporting means and which also serve him as food," declared a 1963 U.S. Fish and Wildlife Service briefing statement on the project. "If the aquatic pastures are inhabited by species which compete directly with sport fishes, little improvement can be expected by stocking with hatchery fish. It is considered good management to remove the existing fishes and to restock preferred species." With a very few notable exceptions, it was a sentiment that was echoed over and over again by officials in all of the agencies involved. It was a paradigm that had been reinforced by generations of fisheries managers at least since the nineteenth century.[48]

As for the more fundamental argument—that the big native fish of the river never could have survived in the Green once the dams were built—it's not so clear cut as Wiley and other defenders of the project have long wanted people to believe. For one thing, nobody had done enough research at the time of the operation to have any idea whether the native fish could survive in reservoirs. In fact, bonytail, the most endangered of the group, survive today behind other dams farther downstream on the Colorado, in places like Lake Mohave and Lake Havasu. The others—Colorado pikeminnow, razorback sucker, humpback chub—have persisted there and near other reservoirs better than anyone could have predicted. Certainly there is a good chance these relatively long-lived fish could have survived in the upper Green at least for a few decades, long enough for another solution to be found.[49]

Whatever the case may be with the natives, the reservoir and the rotenone project have certainly benefited the trout. Between 1963 and 1971, the fish and game agencies responsible for the project had stocked more than 20 million rainbows into Flaming Gorge Reservoir, and they continue to stock these and other fish to this day. The Green River below Flaming Gorge Dam is one of the most stunning trout streams I have ever seen, and from the number of people putting in at the boat ramp below the dam, I'm clearly not the only one who thinks so. Above the dam, I'm told, the rainbows are also doing reasonably well, though the real action these days seems to be in trolling for lake trout and kokanee.[50]

On a larger scale, though, the Green River project was a pyrrhic victory for the rainbows and their proponents. Because of the firestorm that followed, and the environmental movement that the incident helped propel into the national spotlight, no poisonings on such a scale have ever occurred again. In fact, when fish eradication projects are conducted these days, the rainbows are usually the target.

 N I N E

Money Makes
a Way

At about the same time that Miller and Hubbs were feverishly trying to drum up support for native fish and block the Green River poisoning, a backlash of a different sort was also taking place among some of the nation's anglers. These fishermen didn't object to the presence of nonnative trout in their streams—far from it—but they did oppose the catchable hatchery fish that seemingly every fish and game agency was increasingly growing and stocking.

The opposition came largely from the ranks of the fly fishers, who scorned the bait fishermen with their new spinning reels and fiberglass rods and likewise believed the hatchery trout the "worm-drowners" chased were an affront to the sport. They were ugly, with little color and abraded fins from the crowded cement raceways in which they were raised; they were "stupid," having been bred and trained to chase the food pellets thrown at them in the hatchery; and they were tasteless for the same reason.[1]

Instead, these men preferred to fish for "wild" trout, a term, I should point out, that is entirely different from "native." To qualify as wild, in fisheries jargon, a fish need only have been spawned in the wild, a product of natural reproduction. It is a term used only to distinguish those fish that did not originate in a hatchery from those that did. It is entirely possible and even common, in other words, to have a fish that is both wild and nonnative.

As a case in point, it was on Michigan's Au Sable River that the devotees of wild trout found their first rallying point. No trout are native to that stream, only grayling. But by the middle of the twentieth century, introduced brookies, browns, and rainbows had largely taken over, and the river had become one of the premier trout streams in the country. It was there that a group of anglers founded Trout Unlimited in 1959. Their goal was primarily to protect the wild (but nonnative) trout in the river from the hatchery fish that the Michigan Conservation Department routinely stocked.[2]

The motivating force behind the group was an ardent angler named George Griffith, who was also a member of the state Conservation Commission, the politically appointed board in charge of overseeing the department. Griffith had recently won a position as the chairman of the Conservation Commission's Fish Committee, but nevertheless he found it very difficult to slow the department's intensive stocking practices. "In the post-war, plastic rod era," he recalled in his 1993 biography, "everyone had to be able to catch a trout." And so, as his tenure neared an end, Griffith gathered together a group of like-minded and influential men to form Trout Unlimited.[3]

The group quickly attained a broad following. "Letters from anglers across the country seemed to have a common denominator," Griffith later claimed, "a general disgust with their states' management of trout resources." Within a few years, chapters had formed across the United States and even in Canada and Mexico.[4]

But what made Trout Unlimited so attractive in some quarters of society fueled resentment and outrage in others. If, as the group espoused, the stocking of hatchery trout was to be eliminated or drastically curtailed, then other restrictions would also have to follow. For one thing, people who were used to bringing home a string of a dozen trout or more after a day fishing would have to learn to be satisfied with only a tiny fraction of that number. Natural reproduction simply could not produce so many fish, especially with millions of new anglers flocking to the sport. And that meant most, if not all, of the fish they caught would need to be released.

Releasing fish, in turn, meant that worms and other bait were out. Studies then and now have shown that fish swallow baited hooks much more deeply than artificial flies and lures, to the point that it is almost impossible to remove the hook without inflicting a mortal wound. Catch-and-release fishing practiced by bait fishermen would result in a lot of fish floating down the stream belly up.

Such a proposal was guaranteed to be unpopular in certain quarters. But the visionaries and early leaders of Trout Unlimited included men connected to both political parties along with the presidents and CEOs of companies like Rockwell International and American Motors. They were men who usually got what they wanted. "Money," Griffith succinctly declared, "makes a way."[5]

And indeed, Griffith and Trout Unlimited did achieve some early victories. Fly-fishing-only regulations were implemented on sections of the Au Sable and some of the other nicest streams in Michigan early on, thanks to Griffith's efforts as a commissioner. And by 1963, Trout Unlimited had persuaded the Conservation Commission to slash the number of catchable trout it stocked and even to outlaw the keeping of any fish on certain waters—some of the first catch-and-release regulations in the country.[6]

The victories, though, were not without controversy. "You're going to make a violator out of my nine-year-old grandson," one man protested during a hearing on the flies-only regulations on the Au Sable. "He likes to dangle a worm to trout from my dock!" Others roared that "Trout fishing is not just for the rich," and "The working man has just as much right to fish for trout as corporation presidents." One man even quoted the bible, asserting that flies-only regulations would be contrary to tenets laid out by no lesser authority than Jesus Christ himself.[7]

The science of the time also seemed to favor Trout Unlimited's opponents. In one experiment, the Wisconsin Department of Natural Resources closed a section of stream to fishermen to see what effect it would have on the size and number of fish. At the end of five years, there were fewer trout than there had been when the experiment began. So many fish died due to natural causes, they concluded, that anglers might as well keep them; to do otherwise would just be a waste.[8]

The argument went back and forth. Griffith and his allies argued that stocking was a financial boondoggle, since the department's own research had shown that 95 percent of stocked fish died of natural causes within a few months, well before anglers had a chance to catch anything other than a tiny fraction. But the department's hatchery management chief used the same evidence to argue that the state should stock even more often, and with more fish.[9]

The conflict was not easily resolved, largely because there was no evidence that hatchery trout caused any actual harm to the wild fish. Trout Unlimited anglers objected to stockers mostly because they were ugly and lacking in fight or wile, and because hatcheries used money that could be spent on habitat. With such objections, the fish themselves became metaphors. On one side, wild trout anglers derided anyone who fished for hatchery trout as unsporting and unaesthetic. On the other, bait fishermen resented the beautiful people who wanted the rivers reserved for wild trout. In fact, the controversy generated much the same battle that had occurred a century earlier, when wealthy urban elitists had fought rural subsistence fishermen for control of the best waters.[10]

In 1964, though, Michigan's wild trout enthusiasts won

a decisive victory when the Michigan Department of Conservation finally acceded to their demands and stopped stocking catchable trout. It was a stunning turnabout for an agency that had, up to that point, focused much of its efforts on creating put-and-take fisheries. Officially, the department stopped stocking large fish because they cost a lot of money and, though more likely to be caught than fingerlings, still weren't providing a sufficient return to creel. Griffith and his companions, who so despised these "lifeless blobs," also had a lot to do with it.[11]

From there, though, the effort stalled out. Trout Unlimited found it much harder to convince the fish and game departments from other states that stocking should cease, especially because the group had no good evidence to show that stocked fish were doing any harm to wild fish. Most fisheries managers believed that by stocking catchable fish they were just supplementing the wild population and putting "a little cream on top" for the anglers.[12]

WHEN TWENTY-SIX-YEAR-OLD Dick Vincent joined the Montana Department of Fish, Wildlife, and Parks in 1966, he was charged with developing new ways of estimating fish numbers in the state's largest rivers. Seemingly mundane, it was, in fact, a critical job. Up until that time, the only known methods for such work were crude and subjective. And without good estimates, fisheries managers found it almost impossible to know whether their efforts were doing any good, be it through regulations, stocking, or habitat management.[13]

To solve the estimation problem, Vincent didn't try to come up with anything revolutionary. He had finished his master's degree in fisheries only a few weeks earlier, and he thought he could cobble together a workable solution for big rivers based on techniques he had studied in school. First, he had to figure out how to capture the fish. Somebody long ago had discovered that fish can be stunned by electricity. Apply the right voltage and they will float to the surface where they can be easily captured, studied, and eventually returned to the water unharmed. In fact, biologists had long been electrofishing small creeks. They would simply bound a five-hundred-foot section with a couple of nets, stick some probes in the water, and like magic, they could count every fish.

Vincent decided to scale up this technique with more powerful boat-mounted shockers. And instead of using nets, which were nearly impossible to keep in place with so much water, he used statistics—sampling instead of censusing the population. It took a couple of years to work out the kinks, but eventually Vincent grew confident that he could come up with reasonable estimates, and he began looking around for ways to put his newfound abilities to use.

Heading in the nutrient-rich waters of Yellowstone National Park before flowing north through Montana to join the Missouri, the Madison River has always been a very productive stream; it holds an abundance of life from the microscopic organisms at the bottom of the food chain up to the trout at the top. Scenery and plentiful large trout made the river famous among anglers in the 1930s, and by the middle of the century some of the most accessible reaches saw tens of thousands of angler days every year.[14]

To support this intense fishing pressure and the economy of nearby towns, the Montana Department of Fish, Wildlife, and Parks began stocking the river with catchable rainbows in the 1950s. Each year, though, the locals would complain that the fishing wasn't as good as it used to be, and bargain for an even higher quota, until eventually the agency found itself stocking five thousand to ten thousand catchable rainbows per mile in the most popular reaches every year.[15]

Vincent, who had done some of his electroshocking experiments on that stretch, led the effort to determine if and why the fishing was declining. The prime suspect, Vincent told me many years later, was a reservoir upstream. Nobody knew how it was affecting the fish in the river below—it might have been drying up the spawning beds when they needed to be wet, or flooding other critical habitat later in the season—but one thing was clear: the wild fish in the river were having a very difficult time producing any offspring.

So Vincent persuaded the reservoir managers to change the way they filled the reservoir, hoping to make the flows in the river downstream a little more hospitable for the trout, and then used his newly invented monitoring techniques to see what happened in a couple of different sections. To his surprise, though, when they looked at it the next year, only one section had improved. The other one didn't significantly change. "So then we asked, OK, what are the factors that are different?" Vincent recalled. "The only difference we could really see was that one was stocked and one was not. That's the only thing that jumped out at me."

Fish stocking was the primary tool in the chest at that time, and decisions about numbers were usually based on the squeaky-wheel school of management. Anglers would write letters asking for an additional ten or twenty thousand fish in their favorite stretch, local chambers of commerce would lobby their state representatives, and "whoever hollered the loudest got the most fish." (At that time, Montana was stocking almost entirely catchable rainbows.) "The idea then was that the stocked fish were an addition to the wild populations, that two plus two equaled four," Vincent explained. "But a few of us biologists wondered if maybe two plus two equaled three or even less."[16]

A few more experiments over the next few years confirmed it: stocked fish were devastating their wild cousins, wherever they were placed. Stop stocking and the population of wild rainbows would explode by up to 800 percent; the number of wild browns would easily double. And not only that, the wild fish were bigger, more fun to catch, and tastier to eat.[17]

Why did stocked fish have such an impact on the wild fish? Most likely because hatchery fish lack any sense of etiquette. In a crowded raceway where the shadows on the water usually come from attendants bringing dinner, the most aggressive and fearless fish are rewarded. For wild fish, on the other hand, where food is limited and overhead shadows often signal danger, it is a different game. "There's a pecking order. Everybody knows their place, and they don't waste a lot of energy fighting."

Introduce hatchery fish into a river full of wild fish and chaos results. The hatchery fish pick fights with the wild fish, wasting calories on both sides. Many of the wild fish are displaced from their holds, using even more energy. At the same time, the hatchery fish quickly die, either because they are picked off by osprey or angler, or because they use up all of their energy reserves and have trouble finding food. Two plus two, in other words, often equals one.

It was a bombshell of a data set, a discovery that fundamentally changed the terms of the debate. For years, every fish and game department in the country, including Montana's, had preached the gospel of hatcheries, and most anglers had become believers. By the mid 1960s, some fisheries biologists had become skeptical of the program; at the very least, they believed, it was economically indefensible. But so popular had the program become that when Montana's fisheries chief proposed to limit stocking in some of the state's best rivers as Michigan had done, he was quickly shot down by the politically attuned wildlife commission.[18]

Then, along came Vincent and his stunning findings. In 1973, the fisheries chief again went to the commission, this time proposing not just to stop stocking in a few rivers, but to stop it everywhere there were wild fish. Only reservoirs and other waters where natural reproduction was impossible would get the hatchery version. His proposal "resulted in several of the wildest public meetings" the department ever conducted, with irate anglers and business owners mounting personal attacks on Vincent and other members of the department.[19]

The department itself was polarized as well. There were what Vincent calls "some hurt feelings," and "fears" among the hatchery personnel. But, Vincent acknowledged, "I understand that. Here comes some twenty-seven-year old saying 'what you've been doing for the last 30 years is hurting things.' These were lifetime careers. You probably wouldn't be real happy about hearing that."[20]

Vincent sometimes claims he had no agenda, that he just happened on the data and that the numbers spoke for themselves. He described his reaction as "I don't personally have an opinion. It's just, here's the data, what do you think?" But Vincent also acknowledged at least some level of advocacy. In response to those who wanted to continue stocking, "I kept arguing, is Montana known for little fish, or do we want big fish? What really attracts people to Montana? I don't think it's 8-inch fish, because I'll guarantee you that Pennsylvania and New York can stock a hell of a lot more of those than we can."[21]

Dick Vincent is a charming man. He's unafraid to laugh at himself, happily admitting, for example, that before he took a job with the department, he had been one of the squeaky wheels, writing letters asking them to please, please stock more fish in his favorite section of stream. He is the type of guy who can make you feel comfortable and at home, whether he's talking about his grandkids, reminiscing about his early days with the Department of Fish, Wildlife, and Parks, or discussing his latest research.

I'm sure it is thanks as much to his disarming humor and affable manner as it is to his data that Vincent and the wild trout enthusiasts won the debate in Montana in the end. And it all seems to have worked out for the best. None of the hatchery workers lost their jobs—they just changed emphasis. The department began stocking reservoirs more heavily, using smaller, wilder fish, and also propagating natives like the westslope cutthroat. As for the fishing, there were a couple of slow years, Vincent admits, but Montana has since become the fly-fishing Mecca of the world. And Vincent has become something of a legend, at least in the realm of fisheries management.[22]

OF COURSE DISDAIN FOR hatchery trout did not spontaneously materialize in the middle of the twentieth century. Even in the nineteenth century, when fish culture's promoters were at their most vociferous, there were laments. Early on, a select group of anglers complained that the rainbows being introduced to the East were ugly and tasteless, "like black bass from warm and muddy waters."[23]

And once the fisheries agencies began experimenting with raising the trout in hatcheries and stocking them as catchables, anglers complained about that as well. Aldo Leopold, the patron saint of wildlife management in the United States, declared in 1933 that, for sportsmen, "the recreational value of a head of game is inverse to the artificiality of its origin, and hence in a broad way to the intensiveness of the system of game management which produced it." It was a sentiment that was easily translated to hatchery trout. One fisheries manager from Connecticut complained in 1933 that the anglers always "consider a well formed and good colored trout as a 'native' while a poor colored or poorly developed trout is to them a 'liver fed' or 'hatchery trout,'" even though they were frequently wrong about their true origins.[24]

Nor was it just the anglers. A small crowd of scientists also took issue with the practice. "The quick growth of this fish indicates a voracious appetite, which may result in depriving our native species of food," forewarned one prominent critic of rainbow trout as early as 1884. "Like the English sparrow, they may be more easily introduced than banished." In a heated exchange at a meeting of the American Fisheries Society in 1911 another scientist declared that blindly introducing fish into new waters "can no longer be defended as either scientific or practical; it is simply ignorant." In 1931, the dean of the University of Washington School of Fisheries even admitted that "in some sections an almost idolatrous faith in the efficacy of artificial culture of fish for replenishing the ravages of man and animals is manifested, and nothing has done more harm than the prevalence of such an idea." Four decades later, trout biologist Robert Behnke lambasted agency justifications for such programs, declaring that "catchable trout don't meet a demand, they create the demand for more fish."[25]

But although critics continued to issue such rebukes

throughout the early decades of the twentieth century, they were, until the 1960s, almost always on the losing side. Why? For one thing, it seems to have been one of those debates in which each side becomes locked into its position, and cannot back down without losing credibility. Fisheries managers, entire agencies, and even the overseeing commissions had for so many years heralded fish stocking as the panacea for the nation's ills. A whole series of films made by the Department of Interior between the 1920s and the 1960s (and still available for some entertaining viewing at the National Archives in Washington, D.C.) told the public that the federal government was responsible for providing them with catchable fish.[26]

After such a propaganda blitz, managers had a very difficult time acknowledging that they might have been wrong. And since they were ultimately the ones in control of the funds, the hatcheries, and even the fish, the government managers typically won the arguments about policy.[27]

Just what changed this dynamic is unclear. The rapidly increasing use of catchable fish in the years that followed World War II probably played a role. And the fact that the shift occurred in the 1960s and early 1970s, when society itself was in upheaval, probably also provides a clue. Rachel Carson's book *Silent Spring*, in 1962, had profoundly shaken Americans' hitherto unquestioning belief in the benefits of technology. Faith in government also plummeted. In 1964, three out of every four Americans believed the government could be trusted to do the right thing most or all of the time. By 1976, that number had dropped to one in three. In fact, Montana's decision to stop stocking occurred at exactly the same time that the Watergate scandal was reaching its climax in Washington, D.C.[28]

And it's probably no coincidence that the science of ecology was undergoing a seismic shift during this tumultuous era as well. "It was a revolution," recalled one ecologist who attended graduate school during the 1960s. "Prior to the 1950s nature was simplistic and deterministic; after the 1950s nature became complex, fuzzy edged, and probabilistic." In the first decades of the twentieth century, in other words, a majority of ecologists viewed the world as a predictable system. Address a few variables like competition and predation, and one could manipulate the natural world, usually for humanity's benefit. But by the early 1960s, such a position was no longer tenable in ecology or in many other sciences. The world was too complex to ever be completely predictable.[29]

Chaos reigned, in other words, in the streets, in government, in science, and in the natural world. For fisheries managers, the "trust us, we're experts" line no longer worked so well. At a minimum, they could expect some level of public scrutiny wherever and whenever they stocked fish. And that meant hatchery rainbows, the primary product of the hatchery system, would also face new questions in the years to come.

TODAY, TROUT UNLIMITED boasts an annual budget in the tens of millions of dollars and a membership roll of more than one hundred thousand people, including many of the most wealthy and powerful in the United States. Splinter groups and other like-minded clubs swell the numbers even further, making it a constituency that no fish and game agency can afford to ignore.[30]

The group often collaborates with other organizations and agencies and usually emphasizes water quality and habitat issues over fishing regulations, a strategy that has evolved at least in part to forestall charges of elitism. Who, after all, is going to stand up and argue for more polluted drinking water?

But neither does Trout Unlimited shy away from a fight. The group has sued state and federal agencies to protect wild and native fish from multiple threats, including stocked rainbows, and has conducted sophisticated lobbying and public relations campaigns to get its way. In 1997, for example, the group publicly blasted the Colorado Division of Wildlife for its heavy-handed stocking program, generating a public relations battle that lasted for months.[31]

Sometimes the organization seems a little schizophrenic about wild but nonnative fish. In 2004, for example, a Trout Unlimited chapter went to war with a group known as Walleyes Unlimited over who were the rightful denizens of some Montana lakes. Trout Unlimited was concerned that introducing walleye (as their opponents proposed) would harm the wild rainbows that had been introduced decades earlier. At almost the same time, another Trout Unlimited chapter was suing the federal government to list one of the state's native cutthroats as an endangered species. That move, had it been successful, would likely have resulted in the removal of the wild rainbows that were the primary threat to the cutthroats.[32]

Many of Trout Unlimited's battles, in fact, have been internal. And because of its prominence and loose governance structure, they have also been highly public. Questions about Indian and Eskimo fishing rights and about stream access laws, for example, have resulted in rancorous disputes between the national leadership and its local chapters.[33]

Overall, though, the organization that began in Griffith's living room those many years ago has been a remarkable success, especially when it has focused on hatchery trout. The group has never been able to entirely dodge the charges of elitism, but it has moved the anti-stocking movement away from the sport's fringes and into the mainstream.

Montana, for its part, continues to prove that hatcheries are not the panacea that George Perkins Marsh described way back in 1857, that they are not even the unavoidable companion of progress as so many apologists since that time have professed them to be. In fact, Montana has shown what can happen when the technological fix is taken off the table. Montana's lawmakers have done their bit, enacting a powerful set of laws to protect the state's waters and its wild trout from development. Much more importantly, though, the public now feels a strong sense of ownership in its rivers. Take the Madison. When stocking was the norm, not a single person ever questioned the way the dam operators were altering the flows. Today, according to Vincent, "they change the flows in the river by a tiny amount, and people are complaining." And it's the same story on other rivers throughout the state, where dams and other projects have been blocked or altered on behalf of the trout.[34]

As a result, the fishing in Montana ranks as some of the finest in the world. The fish are gorgeous, two-pounders are common, and the rivers spectacular. And you don't often hear complaints from small-town businesses about the lack of stockers either. Montana's residents have become some of the most ardent anglers in the country. Along with the pilgrims who come from out of state to cast a line in these legendary rivers, fish-chasers annually spend somewhere around $500 million in Montana, a healthy sum for a state that claims only about one million residents.[35]

Other states, too, have started emphasizing wild trout. Some, like Wisconsin, began reducing or even eliminating stocking in certain streams at about the same time as Michigan and Montana, largely for economic reasons; there was just no point in pouring money into hatchery fish where their wild cousins were producing plenty of offspring on their own. Eastern states like Pennsylvania followed suit in the 1980s, setting aside hundreds of miles of stream for wild trout. And more recently, Idaho, Wyoming, Oregon, and California have joined the movement, reducing their once prolific stocking programs and reserving thousands of miles of rivers for wild trout. Even states like Colorado, which are often vilified by wild trout advocates as the "rubber fish capital of the world," have felt the need to set aside token waters for wild trout management.[36]

Federal agencies, too, are talking the talk and, in some cases, even walking the walk. In 1963, the National Park Service declared its primary goal would be to maintain and, if necessary, recreate the "biotic associations . . . as nearly as possible in the condition that prevailed when the area was first visited by the white man." By the 1980s, trout stocking had ceased in many national parks, and today the Forest Service, too, faces increasing pressure to halt or at least reevaluate such programs.[37]

The U.S. Fish and Wildlife Service, which took over the U.S. Fish Commission and today manages the National Fish Hatchery program, has also started to deemphasize hatchery trout, though not without considerable controversy. For many years, the agency was the biggest farmer of rainbow trout and other sport fish, producing millions of pounds every year. It was one of those federal programs that few wanted to question. But in 1994, the service's new director, Mollie Beattie, asked a panel of experts to review the national fish hatchery effort. And when they issued their report at the end of that year, the group declared that "the provision of hatchery fish for recreational fishing is not a federal responsibility," and that the Fish and Wildlife Service should instead focus its efforts on things like restoring habitat for native fish.

It was probably the outcome Beattie had been looking for. At least one writer believes she convened the panel to provide political cover when she tried to do what she had planned to do all along: scale back her agency's hatchery program. It didn't quite work out as planned; the sportfishing industry and the hatchery bureaucracy launched a potent counterattack, and Beattie was never able to push through any real reform before she retired in 1996. But the wheels were put in motion, and congressional hearings and a scathing audit from the General Accounting Office have picked up where Beattie left off. And in truth, National Fish Hatchery officials have made some efforts at reform, in some cases by simply giving their hatcheries away to the states, in other cases by trying to retool and propagate declining native fishes. Nevertheless, the U.S. Fish and Wildlife Service is still raising prodigious numbers of sport fish every year, including 8 million rainbow trout weighing almost 2 million pounds.[38]

In short, the struggles at the U.S. Fish and Wildlife Service mirror the struggles taking place among fisheries managers, anglers, and groups like Trout Unlimited all over the continent. The wild trout movement has made tremendous gains over the past few decades. Native trout are also gaining a constituency. Nevertheless, catchable rainbows are at little risk of disappearing any time soon.

 T E N

The Way of the
Passenger Pigeon

y 1991, Dick Vincent had been working at the Montana Department of Fish, Wildlife, and Parks for twenty-five years. The man who had begun his career challenging the established wisdom regarding hatcheries had become the establishment—as the fisheries manager for some of the state's most famous waters, rivers like the Yellowstone, the Madison, the Missouri, the Big Hole, and the Gallatin. He spent most of his time behind his desk or conducting meetings and a lot less time in the field than he used to. His first hint that a crisis was occurring, therefore, came in the form of a few numbers on a piece of paper.

Earlier that summer, several of the department's biologists had surveyed the fish in the Madison using the electrofishing techniques Vincent had pioneered decades before. They had found plenty of lunkers—more than usual, in fact—but almost no one-year-olds. An entire generation of fish seemed to have simply vanished, if they had ever existed in the first place.

Vincent's first reaction was that the biologists were doing something wrong. But they knew what they were doing, and Vincent trusted them. He wrote it off as one of those vicissitudes of nature that people were not always privileged to understand. He figured he'd keep an eye on it and see what happened next year.

Three years later, Vincent finally faced up to the cold truth. The fishery on the Madison, a fishery renowned the world over, a fishery very close to Vincent personally, was crashing. And he had no idea why.

Vincent had heard that Colorado was experiencing a similar problem and that a man named Barry Nehring was heading the investigation. The two men had known each other for many years; they first met to exchange ideas on fish estimation techniques decades earlier, and they had continued to renew the relationship at conferences and meetings in the years that followed. Vincent picked up the phone.

WHIRLING DISEASE FIRST grabbed attention more than a century ago, when it struck the rainbow and brook trout that fish farmers in Germany were trying to raise alongside the native browns. These American fish had first arrived in Germany in the late 1870s and early 1880s, a gift of the U.S. Fish Commission. Observers at the time described misshapen fish, often with gnarled black tails, and a propensity to whirl around in zany corkscrew patterns when disturbed. The fish would continue circling until exhausted, or "unscrew themselves right onto the banks." Sunken eyes and twisted jaws gave the fish a particularly ghastly appearance. And when pathologists examined the carcasses under a microscope they found lesions full of a tiny parasite that had also destroyed the cartilage of the young fish. They gave the animal a Latin name, *Myxobolus cerebralis*.[1]

If you imagine a microscopic grappling hook, you have some idea of what *M. cerebralis* looks like right before it infects a fish. Upon contact with its prey, three coiled springs discharge

Under a microscope, a *Myxobolus cerebralis* triactinomyxon looks like a grappling hook. At this stage, the parasite is ready to attach to a fish. When it does, three coiled springs in the tip (the dark portion on the right) shoot into the skin, providing a secure entrance route for the germ capsule.

Vicki Blazer, U.S. Geological Survey

into the skin, holding on to the fish and making a tunnel for a germ capsule known as a sporoplasm to enter the flesh. Once there, the invader ejects its contents: sixty-four germ cells that multiply and burrow through the fish until they find cartilage, with the skull and spinal cord being especially attractive. Once there, the cells develop into a second stage known as a myxo-spore—another capsule full of germs—and wait for the fish to die. And when it does, if the fish is consumed by an aquatic species of worm known as *Tubifex tubifex*, the parasite can complete its life cycle. New grappling hooks, new fish, new myxo-spores, new worms.[2]

It might seem that such a complex life cycle, dependent on two hosts, would be vulnerable to disruption, and that the parasite would be easy to eradicate. It's not. Tubifex worms thrive in waters throughout the world, and the parasite itself is incredibly hardy. Freeze them, dry them out, do what you will; spores resting in the mud remain viable and ready to infect the next target for thirty years or more.[3]

What's more, of all the trout and salmon, the rainbows

are far and away the most susceptible to whirling disease. The browns, native to Europe, can carry the parasite, but seldom experience any ill effects, probably because the parasite is native to that continent as well. Having evolved alongside it for thousands of years, the browns developed some sort of resistance. Not so the naive Americans.

That's the way we understand the disease now. But until the middle of the 1980s, very little of the parasite's life cycle was known. All the nineteenth-century European culturists knew was that a nasty microscopic organism was killing and deforming many of the rainbow trout around which they had developed an industry.

It wasn't Armageddon. Hatchery managers found that they could mitigate the effects of the disease by raising the youngest fish in small tanks of spore-free well water, putting them into contaminated ponds only when they were larger and much of their cartilage had hardened into bone. And they learned to live with some loss as a cost of doing business. It was a flaw with the fish, but not enough to outweigh the relative ease with which rainbows could be raised in hatcheries. In fact, it is a testament to just how popular rainbows had already become that the Europeans continued to culture them despite their susceptibility to this disease.[4]

M. cerebralis crossed the Atlantic sometime in the 1950s. Some people like to blame an infected package of frozen trout that may have been ground up and fed to fish in a hatchery in Pennsylvania. But in truth, nobody really knows how it got here, and it seems likely that with the rise in global commerce that followed World War II, it could have made the trip multiple times.[5]

Whatever the case may be, by 1970 whirling disease had invaded hatcheries on both coasts and along the Great Lakes. Having grown increasingly dependent on rainbows, and with no good methods for decontaminating hatcheries, fisheries managers feared the apocalypse was upon them. Infected fish were destroyed and buried in quicklime, hatcheries quarantined and sometimes even razed.[6]

Most infected rainbow trout die from whirling disease at a very young age. Those that do survive usually have deformed skeletons and skulls, bulging eyes, and black tails, like these.

S. L. Hallett

After a while, though, such draconian measures began to seem a little over the top. Good nutrition, the relatively benign environment, and a combination of chemicals seemed to keep the disease in check in the hatcheries, and after a while, many culturists just shrugged their shoulders. "The trout have learned to live with it," said one Pennsylvania fish commissioner as late as 1995.[7]

And for a couple of decades, the disease map didn't change much either. Hatcheries in a broad swath down the middle of the country remained curiously unaffected, and most hatchery managers believed it was under control. Whirling disease simply joined a long list of other maladies, many with much more serious consequences, or at least more sinister names, like viral hemorrhagic septicemia or infectious pancreatic necrosis.[8]

As for the field biologists, whirling disease wasn't even on their radar screen. They may have heard of it, perhaps from a hatcheries management class in graduate school, but few could have dredged up the Latin name of the parasite that caused it. Everybody knew that whirling disease simply wasn't a problem in the wild.[9]

A DOZEN MILES BELOW its headwaters in Rocky Mountain National Park, the Colorado River flows west for a dozen miles through a gentle valley of ideal ranchland. It's not the coursing torrent that it is downstream; it meanders gently, and in many places the riffles are easily waded. Tall cottonwoods provide dappled shade while they are standing and ideal cover for the fish once they fall.

Introduced sometime in the early part of the century, the rainbows from this section of river were renowned among fisheries managers for both their longevity and their hardiness, and for many years, these Colorado River rainbows, as the strain is known, were one of the keystones of Colorado's fisheries.[10]

In the early 1990s, this section was also a terrific place to fish. Brown trout were plentiful, but it was the rainbows that really made the place. In some sections, a thousand rainbows occupied every mile of stream. That's pretty good in terms of numbers, but what made the area so amazing was that nine hundred of them weighed a pound and a half or more.[11]

For Colorado Division of Wildlife biologist Barry Nehring, though, those numbers represented catastrophe. Nehring had been studying that stretch of river for much of his fifteen-year career with the department. It was he who had directed the electrofishing that yielded the fish population estimates. And what jumped out at Nehring in the fall of 1993 was not how many big rainbows there were, but how very few small ones — the same thing that was puzzling Vincent on the Madison.

It was a confusing pattern. Browns were reasonably abundant across the entire size spectrum. And plenty of the very youngest rainbows, the ones that had hatched that year, swam among the shallow nursery habitat along the banks. But the one-and-a-half and two-and-a-half-year-old rainbows seemed to have completely disappeared.

The light-bulb moment came after Nehring returned to his office and called his boss to hash things out. Almost casually, aquatic research chief Tom Powell threw whirling disease into the conversation. It fit the pattern: it devastated rainbows but seldom browns, and it often killed the young fish a few months after they hatched. And sure enough, when he returned to the river a month later, almost all the rainbow fry Nehring had observed on his earlier trip were gone. For three days, a Colorado Division of Wildlife crew electrofished twenty miles of river, and they were only able to find four rainbow fry. It was as though, he later recalled, "someone had just vacuumed them all out of the bottom of the stream."

One of the fry that they did find had the "frog-eyed" appearance characteristic of trout with whirling disease, and lab tests confirmed it. The young fish had *M. cerebralis* cysts in their skulls. The infection was light, but that was to be expected. All the fish with the heavy infections, which is to say most of them, would have died and decomposed already.[12]

A month later, Nehring issued a memo, a document that would ultimately shake the department to its very core. Nehring knew the memo would face a skeptical and potentially even hostile reception, and much of the document reads like a legal brief, with the evidence painstakingly laid out and every possible counterargument carefully rebutted. In the end, though, Nehring allowed himself one rhetorical flourish. If these die-offs continue, he declared, "We have a 'grace period' of only 1–2 years before the Colorado River rainbow trout population . . . goes the way of the passenger pigeon." The old fish would begin to die, and nothing would be there to replace them.[13]

COLORADO FIRST DISCOVERED that whirling disease had invaded the state in the fall and winter of 1987 and 1988, when tests showed the parasite had contaminated one of its biggest hatcheries, along with several private operations. In a panic, the state initially quarantined the hatcheries and the Division of Wildlife held an emergency meeting on the matter. During the meeting, though, department officials were reassured to learn that in the decades since it had arrived in the United States, nobody had ever reported problems from whirling disease in the wild. The leaders of the department concluded that the panic was misplaced and lifted the quarantine, and that summer the department began stocking the infected fish into rivers and lakes throughout the state.[14]

On one hand, it was an understandable decision, given the information available at the time. But on the other hand, Colorado was remarkably cavalier about the disease compared with its neighboring Rocky Mountain states, most of which staunchly refused to stock fish from any hatchery where the *M. cerebralis* had been detected. Colorado officials, though, continued to release such fish throughout the state.[15]

I don't know what agency officials said, in private, or what they thought, after they read Nehring's memo those many years ago. Did they understand the scourge they'd released upon the state by stocking infected fish? Were they sickened? Defensive? Whatever their private thoughts may have been, there weren't a lot of public mea culpas. As soon as the press picked up the story, Division of Wildlife officials accused Nehring of "spreading unnecessary panic," and lambasted reporters for being "alarmists," "reactionary Chicken Littles," and "headline-happy sensationalists with an overdose of imagination and a shortage of facts."[16]

Unapologetic, the higher-ups in the agency decided to continue stocking "lightly-infected" fish into waters that already held the parasite (not a great sacrifice, since by the mid-1990s whirling disease had contaminated thirteen of the state's fifteen major river drainages). When neighboring states objected, fearful that the disease might not respect political boundaries, Colorado officials brushed them off. When Colorado state senators urged a stop to the practice, even proposing a nonbinding resolution to that effect, the department leaned on its allies in the legislature and sent the bill packing. And when Trout Unlimited issued a press release comparing the stocking of "lightly infected" fish to the selling of hamburgers that were "lightly contaminated" with *E. coli,* one of the department's commissioners blasted the group for what he called a "terrorist tactic."[17]

The mindset was simple: "either we stock fish or lose a tremendous opportunity." The fisheries would collapse and the anglers would disappear (not to mention the fishing licenses, the department's main source of revenue). Department officials simply could not imagine a world without hatcheries.[18]

By 1999, Nehring's research had showed that stocking infected fish was never a good idea. Even if the watershed had already been contaminated with the parasite, continued stocking of such trout dramatically increased the prevalence of the disease, both where the stocking occurred and for miles downstream. But officials still stalled. It was not until 2003, in the face of overwhelming evidence, and after spending well over $10 million to decontaminate only some of its facilities, that Colorado finally stopped stocking fish from hatcheries infected by the *M. cerebralis* parasite. By that time, though, it was too late. The disease had established itself in the wild, and the department's policy of stocking diseased fish, Nehring later declared, was the primary cause.[19]

I FIRST MET BARRY NEHRING at the bottom of a spectacular desert canyon on Colorado's western slope known as the Gunnison Gorge. I'd hiked down to the site on a crisp fall morning in 2006 and found a camp full of dozens of Division of Wildlife personnel preparing to electrofish and survey the river. A wiry man of medium build, Nehring had a sort of magnetic and disciplined energy that made him a locus of the buzzing activity; he seemed to have no problem filling me in on whirling disease while at the same time directing the electrofishing crew, identifying and measuring fish, and even discussing the menu for the evening.

The fishing on the Gunnison River in this section was, in the early 1990s, even more remarkable than the stretch of the

Barry Nehring continues to monitor the Gunnison Gorge in Colorado, where he discovered that whirling disease had destroyed the rainbow trout fishery in the 1990s.

upper Colorado where Nehring first discovered whirling disease. In the best sections, there were ten thousand fish per mile, about half rainbows and half browns, and it was easy to catch fish weighing two pounds or more. But then, in 1993, Colorado Division of Wildlife officials decided to stock fish they knew to be infected with *M. cerebralis* into the Blue Mesa Reservoir, fifty miles upstream, and by 1994, the river through the Gunnison Gorge was contaminated. The disease had inflicted a 99 percent mortality rate on young-of-the-year rainbows, Nehring said, every year since.[20]

Because they are so much more resistant to it, whirling disease actually works to the advantage of the brown trout. It's almost a weapon. The browns may still carry and spread the parasite, but they are seldom killed by it. As a result, in many of the rivers stung by whirling disease, the brown trout take over. A decade after the disaster struck, in fact, there were just as many trout in the Gunnison Gorge as there had been beforehand, except that by that time they were all browns.[21]

When I asked Nehring why this was a problem from a fishing point of view, he nodded his head toward some anglers on the other side of the river, a group I'd been eyeing since I arrived. Every one of them, it seemed, had a personal guide, but I hadn't seen anyone catch anything yet. Nehring estimated that every one of them had paid two or three thousand dollars for three days on the river, and he seemed to think that if they could afford that, they probably had careers that prevented them from spending every day fishing.

Rainbows launched to fame a century ago in large part because they are such aggressive fish. This means not only that they put up a good fight once hooked, but that they are far more likely than the wary browns to take a hook in the first place, especially when that hook is doing a poor job of pretending to be a fly—especially, in other words, when there is an inexperienced angler at the other end of the line. Nehring estimates that a rainbow is three to ten times more likely to be caught than a brown. Use the big number and assume the fishing is largely catch and release. This means that fishing a section of river with ten thousand browns is about as good as fishing another section of river with only one thousand rainbows. It's small wonder that Colorado wanted the rainbows back.

And so the story goes full circle one more time. Loop back to the nineteenth century, when the U.S. Fish Commission was busy shipping rainbow trout around the country and around the world. Some of those fish eventually made it to Germany and into the hands of an aquaculturist named Hofer. Many of his rainbows became sick with whirling disease, but some survived. And generation after generation, natural selection improved their resistance, until in the early part of the twenty-first century, an Egyptian fish biologist and professor at the University of Munich almost accidentally discovered that they were nearly invulnerable to the parasite. News travels fast, and by 2004, Colorado Division of Wildlife officials were spawning "Hofer" rainbows with some of their own Colorado River rainbows, crossing their fingers, and releasing the progeny into select waters, including the Gunnison Gorge.[22]

While I was there, the electrofishing crew caught a few of the stocked Hofer hybrids, and they looked healthy to me. But since they were still too young to spawn, the real question remained unanswered: whether these Hofer hybrids would successfully produce offspring in the wild that were resistant to the disease. A year and a half later, though, I received an ecstatic email from the Division of Wildlife. The crew had returned to the gorge in the fall of 2007, captured young-of-the-year rainbows, and through DNA testing discovered that the experiment, so far at least, had been a success. Wild rainbows seemed to be on the verge of a comeback. And, the email concluded, the department planned to release millions more Hofer hybrids across the state beginning in 2008.

Not everyone, of course, will be cheering. Some biologists have publicly derided the idea that Hofer rainbows would make worthy progenitors for Colorado's trout. These are fish, they point out, that have been bred for over a century by commercial trout farmers. Traits that served them well in the hatchery—fearlessness, intensive calorie-burning aggressiveness—will serve them poorly in the wild where predators are abundant and food is scarce. Even worse, they are likely to interbreed and thus "dilute and degrade" the genetics of those wild rainbows that have managed to survive the epizootic (as epidemics among animals are called).[23]

Others point out that in an age of globalization and climate change, whirling disease is the tip of the iceberg. Untold new diseases are just waiting for an unwary angler to bring them home from some exotic locale on his unwashed waders and release them into America's rivers. The best defense we have against such an attack is a diversity of fishes, one of which may be resistant. If, on the other hand, the entire population of trout were to be descended from one small raceway of Hofers, and if those Hofers happen to be susceptible, the effects could be even worse than what happened with whirling disease. Some go so far as to say it's just a matter of time.[24]

AS FOR MONTANA, the state began testing the fish from the Madison for whirling disease almost as soon as Dick Vincent got off the phone with Barry Nehring. Sure enough, they were infected. For some time thereafter, an atmosphere of complete panic pervaded the department. "We didn't have any data, so we basically assumed the sky was falling," Vincent recalled. "We thought everything was going to be bad."[25]

And in some rivers, the sky did fall. In the Madison the rainbow population virtually disappeared within a couple of years. But in others, even though they were infected, the rainbows hung on.[26]

What caused some rainbow populations to crash while others persevered? Why did Montana (which had largely ceased stocking fish) and Colorado get hit so hard, while other states have yet to see any serious impacts in the wild? Those are the questions that the federal government has spent millions of dollars trying to answer, the questions that hundreds of researchers are still trying to figure out. Vincent cites a multitude of possible factors: the productivity of the stream, the temperature, the flows, the location where spawning occurs, the time at which the trout hatch, the type of *Tubifex tubifex* worms that happen to be present—the list goes on and on. Change any one of these variables, and whirling disease may shift from catastrophic to merely annoying, as so many other states report.[27]

But when I asked Vincent what Montana planned to do about the disease and specifically whether there were any plans to introduce resistant fish, as Colorado had done, he demurred. "I'm a little reluctant to just start whaling around out there, personally," he admitted. "I'm somewhat leery that it may backfire on us."

My meeting with Vincent occurred shortly after I received

the announcement from the Colorado Division of Wildlife about the success of its Hofer rainbow program, and the contrast was striking. The Colorado worldview seemed to be, at its heart, based on the idea that science and technology make it possible for humans to reliably manipulate the natural world to their own advantage. Vincent, on the other hand, seemed to have internalized the idea that nature is inherently unpredictable and that any attempts to manipulate it stand a good chance of going astray. He wasn't an ideologue, just a practical man who's seen Murphy's Law in action enough to know that sometimes it's best to just let the system take its own course.[28]

And in fact, it appears that Vincent's strategy may be bearing fruit. For a number of years between 1998 and 2007, Vincent's team collected some of the few remaining rainbows from the Madison, brought them into the hatchery, spawned them, and tested to see how resistant the offspring were to the parasite. The results were unambiguous. The offspring of the fish from the later years were significantly more resistant than the offspring in 1998. Vincent believes, in other words, that natural selection is breeding resistant trout in the Madison without any help from humans.

In the river itself, the numbers of young rainbows are approaching and in some cases even exceeding historical norms. Older rainbows are still lacking, because at this point the trout seem to be dying off between the ages of two and three years old. Perhaps, Vincent speculates, this is a difficult transition for them as they enter sexual maturity and move into new habitat, and the additional stress finally causes them to succumb to the whirling disease infections they have had since they were youngsters. But give it time, and eventually natural selection may even find a way around this obstacle, repopulating the Madison with naturally parasite-resistant fish.

IN THE END, probably the most lasting consequences of whirling disease will come not through its actual physical impacts on the fish themselves, but through its effects on people: anglers, managers, critics. In Colorado, for example, as whirling disease ripped through the hatcheries and the state tightened its regulations on where infected fish could be stocked, the number of uninfected catchable rainbows available for stocking into many of the best streams plummeted to a tenth of its former level. To preserve what remained, the Wildlife Commission (a politically appointed body that oversees the department) reluctantly agreed to reduce the creel limit on many of the most popular rivers. Anglers, who at one time had been able to fill their freezer with the stocked rainbows they'd caught on a day's fishing trip, were now breaking the law if they possessed more than four fish, or in some cases even two. Whirling disease accomplished, in other words, what Trout Unlimited had been trying to do for so many years.[29]

The result, as Colorado's managers predicted, was a concomitant decline in fishing license sales. In one of the fastest-growing states in the nation, there were fewer anglers in 2001 than there were even five years earlier, the decline being particularly apparent in one segment of the fishing community. "In case you haven't noticed, it now ranks as a relative rarity to find someone, particularly an adult, fishing with bait on moving water," wrote Colorado's best-known fishing columnist in 2003. "The once-common scene whereby someone parks a lawn chair and a cooler streamside while soaking a juicy night crawler in hopes that a trout swims by now almost ranks as an occasion to stop for a photo."[30]

The Colorado Division of Wildlife, too, was profoundly changed by its collision with the disease. Until the early 1990s, Colorado stocked more rainbows on a per capita basis than any other state in the country. Frequently derided as the "Hatcheries 'R' Us" state, Colorado defended its heavy stocking as the only possible means of maintaining the fisheries in waters that, for whatever reason, lacked the ability to produce enough fish on their own. "These people aren't willing to look at the biological basis for what we do," said the aquatics chief for the department,

Eddie Kochman, in 1998. Kochman initially believed the department had no choice but to continue stocking fish infected with *M. cerebralis*.[31]

But as the evidence piled up that the stocking policy had dramatically worsened the epizootic, Kochman did a dramatic and public about-face. He admitted to a reporter in 2001 that the decision to continue stocking whirling-disease-positive fish had been a "big, big mistake." And upon his retirement a year later, he simply declared that "for a long time, I thought we could satisfy most of our needs simply by producing more hatchery trout. I was wrong." The costs of relying on hatchery fish, he declared, had become too great.[32]

Just what the future holds for Colorado's trout remains to be seen. If my conversations are any indication, tensions still simmer within the Colorado Division of Wildlife over the ban on stocking infected fish into waters that might be considered natural trout habitat. And if the department is successful in breeding whirling-disease-resistant trout, its hatchery rainbow program could come roaring back even bigger than before—at least until the next disease strikes. Only time will tell.[33]

A Single New
Mongrel Species

T he Environmental Protection Agency has the responsibility for implementing the Clean Water Act. Thirty years
after the passage of the law, though, the EPA found it still could
not reliably answer some of the most basic questions about the
health of the nation's freshwaters. So at the dawn of the new
millennium, the agency began an ambitious program to determine the condition of the nation's rivers and lakes. The officials
in charge of the project used software to randomly select hundreds of sites throughout the country and then sent teams to
survey them in every state.[1]

Thus, for five years, Amy Ackerman and her partner spent
spring, summer, and fall trailing a GPS around Colorado, trying to find the sections of stream the computer had randomly
assigned. Once they found the site, they followed an exhaustive
survey protocol designed by the EPA. They took water samples
and assessed things like the chemistry, the morphology, the vegetation, the invertebrate population, and, of course, the fish.[2]

They sampled more than one hundred sites, from icy mountain creeks to the warm and sluggish channels of the plains. They endured sun and hail. And once the field season closed, they dutifully returned to the office, cleaned up their muddy field notes, and sent them to the central EPA office. There, scientists scanned them into a computer and analyzed the data. Distilled into a few crisp numbers on a monitor, the results were stunning. Two out of every three individual fish swimming in Colorado were nonnatives. In many of the mountain streams, according to Ackerman, that ratio was much higher.

Colorado is exceptional, but other states like Arizona and Washington had ratios of nonnatives to natives that were almost as high. Considered in terms of the number of species as a whole, instead of individual fish, the statistics are equally dramatic. At least 536 nonnative species now thrive in the freshwaters of the United States, and they dominate some river systems, especially in the American West. The Colorado River, once home to 51 native species, has been swamped by at least 100 nonnative species. They include carp that Spencer Baird introduced from Germany in the nineteenth century. They include small minnows and shiners that arrived in somebody's bait bucket many years ago. And they include sport fish like the basses, the sunfishes, the brook trout, the brown trout, and, of course, the rainbows.[3]

The introductions are not without cost, either. When it last updated the list in 2008, the American Fisheries Society considered 700 taxa (species, subspecies, or unique populations) of North American fish to be endangered, threatened, or of special concern—about 40 percent of the total. And that doesn't include the 61 fishes that slipped into extinction in the twentieth century. In fact, freshwater fishes are some of the most seriously threatened vertebrates in the world, second only to amphibians.[4]

A single cause can seldom be fingered; habitat loss and pollution are frequently to blame. But so, too, are introduced fishes. By some estimates, the introduction of nonnative species has been at least partially responsible for the extinction of 27 species

of North American fishes alone. And hatchery rainbows are perhaps the guiltiest of them all.[5]

How do they do it? Some effects are obvious. Rainbow trout
are voracious predators who occasionally eat other fish. More
importantly, though, they are tough competitors. Dumping a
bunch of hatchery trout into a stream is like putting a school bus
full of teenage boys in front of an all-you-can-eat buffet. There's
not a lot of food left by the time the other customers get to the
counter.[6]

What's more, as Dick Vincent showed in Montana, hatchery rainbows are particularly ill-behaved. Bred and trained to be
aggressive and fearless in the hatcheries, they disrupt the established social order among wild trout, causing them to waste
energy and expose themselves to predators.

Other effects are less obvious. Disease can run rampant
under the crowded hatchery conditions. And while the hatchery
fish may survive thanks in part to the drugs the managers pour
into the water, they carry these diseases with them when they
are introduced to the wild.[7]

Hatchery fish can also harm native fish because they have an
uncanny ability to direct human behavior. For one thing, hatchery fish draw anglers to the area in which they are stocked, and
the wild fish fall prey to many more hooks than they otherwise
would. For another, hatchery fish can hide the deeper problems
affecting an ecosystem, as they have since the nineteenth century. People are much less likely to be concerned about human
impacts if they think there is a technological fix.[8]

In some cases, hatchery rainbows can even lead humans
to modify entire river systems. Enterprising anglers and state
and federal agencies commonly build small ponds or impound-
ments on both private and public land on their behalf. By one
estimate, people built 5 million small ponds in the United States
in the twentieth century, many at least partially for the sake of
the fishing. Even the big dams (about twelve hundred of them)
were built in part for the fishing opportunities they created. In
the rush to build dams across the United States in the boom years

that followed World War II, the Bureau of Reclamation and the U.S. Army Corps of Engineers often produced cost-benefit analyses to justify the projects. And for many of these projects, the increased recreational angling for hatchery rainbows and other fish was a prominent part of the plus column.[9]

But perhaps one of the most insidious effects of hatchery rainbows, and also one of the most fascinating from a philosophical, ecological, and legal point of view, comes not from any sort of antagonistic behavior. It stems from an act that, when humans do it, is called making love.

FIRST DESCRIBED BY Captain Meriwether Lewis on his way to the Pacific in the summer of 1805, westslope cutthroat trout once dominated the waters of the northern Rocky Mountains. At that time, the subspecies swam in river systems from present-day Yellowstone National Park into British Columbia and Alberta. Their range included most of what is today western Montana and northern Idaho as well as unconnected outposts closer to the Pacific.[10]

As discussed in Chapter 5, the ancestors of these trout first arrived in the area one to two million years ago, after swimming up the Columbia River from the Pacific. Some time later, impassable waterfalls closed the gates behind them, allowing the fish to evolve in isolation for eons. More recently, with the aid of the ice dams and shifting watersheds that occurred during the last ice age, a few of these trout even managed to cross the Continental Divide and establish another population in the Missouri River watershed, where an angler with the Lewis and Clark expedition caught them.[11]

Never attaining great size, westslope cutts can nevertheless be quite stunning; they have deep black spots, a red slash on their throat, and, in some spawning males, a brilliant crimson and orange belly. I've never eaten one, but according to Captain Lewis, they can also be "sumptuous" when cooked over an open fire and consumed alongside buffalo humps, tongues, and marrow bones.[12]

Before Lewis, these trout likely nourished local Indian tribes, and they also provided food and entertainment to white explorers and settlers in the years that followed. Ultimately though, settlement, industry, fishing, and introduced species all took their toll, and as the millennium approached, these native trout continued to swim in only a fraction of their former range. While some experts claim the number is as small as 2 percent, others, including biologists with the U.S. Fish and Wildlife Service, estimate that westslope cutthroats occupy 20 percent of the rivers and lakes to which they are native—a discrepancy that largely depends on just what should be categorized as a westslope cutthroat.[13]

Consider the North Fork of the Flathead River, which runs from British Columbia into northwestern Montana next to Glacier National Park. The geology of the area makes the river a difficult place for a fish to find a living. The rock through which the tributaries flow lacks in certain nutrients, and there is consequently very little food to go around. Nevertheless, westslope cutthroat called the Flathead home, the only members of their genus to do so for tens of thousands of years.[14]

Then, at the dawn of the twentieth century, rainbows were introduced to the watershed, many of them direct descendants of fish Livingston Stone and his crew had captured on the McCloud River a couple of decades earlier. But even though fisheries managers continued to pour reinforcements into the Flathead until the state of Montana ceased stocking that river in 1969, the rainbows never seemed to do very well in the North Fork. They certainly never managed to evict the cutthroats from the high mountain tributaries, perhaps due to the unproductive nature of the water, perhaps because they were not well adapted to the temperature or the flows. And so, when stocking ceased, the few rainbows that had made their way into the North Fork seemed largely to disappear.[15]

Rainbow trout, though, are closely related to cutthroats, and the two species can still successfully interbreed and produce fertile offspring. When their progeny then produce their own children, it can be impossible to differentiate between purebred fish and hybrids from looks alone.

In 1998, therefore, a coalition of Montana biologists from various organizations began taking a closer look at the trout in the North Fork of the Flathead and in other putative westslope strongholds. Instead of relying on appearance, they decided to use a different method to identify the fish. They decided to look at their DNA. Over the course of several years, they captured hundreds of fish from throughout the watershed, clipped a small amount of tissue from their fins (releasing the fish alive), and brought the samples back to the lab for analysis.[16]

Because they are so closely related, the genomes of rainbow trout and cutthroats are largely identical. However, with enough work, much of it trial and error, it is possible to find sections of DNA, or markers, that are not. Suppose, for example, you examine the same fragment of the same chromosome in hundreds of rainbow and cutthroat trout. You may have no idea what, if any, purpose that fragment serves. But if, in all of the rainbows, that fragment is 70 units long and in all of the cutthroats it is 153, you can be fairly certain that you have found a diagnostic marker. Now, from the smallest amount of tissue, you can determine what kind of fish it came from. Does it have 70 units? Must be a rainbow. Does it have 153? Cutthroat.

Most analyses use multiple markers—usually at least five. Not only does this make for a greater degree of certainty, but multiple markers also make it possible to estimate the amount of hybridization in those species that are capable of crossing the line. Suppose with your fish, for example, that one of four grandparents was a rainbow and the other three were cutthroats. You'd expect a quarter of the markers to be characteristic of rainbows, and three quarters for cutthroats. Other scenarios can lead to other percentages. And if you use multiple markers on multiple fish, you can estimate the degree of hybridization in the population as a whole. You can ask, in other words, what percentage of the gene pool comes from rainbows, and what percentage comes from cutthroats.[17]

Analyzing DNA in this manner is a laborious process. The researchers on the Flathead spent three years catching hundreds of fish from forty-two different sites. With every sample, they had to grind the tissue and extract the DNA. They had to set up thousands of chemical reactions to amplify and dye the fragments they were interested in, and they had to pay for elaborate machines costing hundreds of thousands of dollars just to measure the length of the markers. But when, after five years, they were done analyzing all the markers on all the fish, they had thousands of data points. And when those data had been distilled into a few numbers, they had some very powerful and alarming results.

For one thing, many of the fish that had looked like pure westslope cutthroats were not. They were, instead, rainbow-cutthroat hybrids. (In some cases they were hybrids between westslope cutthroats and Yellowstone cutthroats, another fish that had been introduced to the watershed many years before.) The extent of the hybridization varied. The fish from sites farthest downstream tended to have a high percentage of rainbow genes, probably because they were closest to Flathead Lake, where rainbows still lived. The fish from some of the high mountain tributaries were mostly descended from westslope cutthroats. But many of them still had a small amount, sometimes as small as one percent, of rainbow ancestry. And that meant the North Fork was not the sanctuary for purebred westslope cutthroats that researchers had believed since they began studying the problem in the 1930s.[18]

Even more troubling, though, was the discovery of hybrids in seven tributaries that had held only pure westslope cutthroats in the early 1980s, when Montana biologists had first used molecular techniques to examine the problem. Very few of the newfound hybrids were the result of purebred rainbows mating with purebred westslope cutts. Rainbows were, after all,

A westslope cutthroat trout (or is it?) found its way into my net on the North Fork of the Flathead River in Montana. Hybridization with rainbow trout has made it difficult to identify or even define this species.

virtually nonexistent in the North Fork. Instead, it appeared that hybrids were begetting more hybrids, spreading this genetic pollution at a very rapid clip.

Why did it happen so quickly? No one is quite sure. One possibility is that trout with only a small percentage of rainbow trout genes were surviving, even thriving, in waters where pure rainbows could not. Mix it with another hypothesis: that hybrid fish are more likely to migrate than purebred cutts, perhaps due to their rainbow ancestry, and you've got a recipe for just such a result.[19]

It's not limited to that watershed, either. Rainbow trout genes have infiltrated most other subspecies of cutthroat trout in the American West, their tendrils growing every year. Some fish, like the Alvord cutthroat, have already gone extinct due to such hybridization. Others, including California's fabled golden trout, are dangerously close. Golden trout were once native only to a small part of the Kern River drainage, but the California Department of Fish and Game sought to increase their range in the 1950s and 1960s by stocking them into new waters throughout the state. They also stocked them back into the Kern River drainage, trying to increase their numbers for the anglers. The fish they stocked, though, were actually golden-rainbow hybrids. The hybrids and the pure goldens interbred, and when researchers went looking for pure golden trout four decades later, they found only a few small maybes—populations so meager that even if they are pure, and even if humans intervene, they are likely to go extinct sometime in the near future.[20]

It's a disturbing and, at the same time, compelling image. Rainbow genes have become their own entity, disconnected from the fish in which they began. And with the aid of human technology they are moving up the North Fork of the Flathead, up the Kern River, and into watersheds throughout the American West.

FRED ALLENDORF'S LABORATORY at the University of Montana is just the sort of place you'd expect to find a man who has been studying trout genetics for the past thirty-five years, a man at the top of his game. Pyrex beakers and boxes of laboratory supplies line the shelves, state-of-the-art DNA sequencing machines and other equipment purr on the counters, and graduate students attending to unfinished research lend just the right amount of chaos and energy.

Allendorf himself does most of his work in a spare and orderly office that, when I visited, was lit only by natural light. Two computer monitors in standby mode displayed slideshows of some of the beautiful places Allendorf has visited for his research, and a Buddhist-sounding gong occasionally made it known that a new email had arrived.

At the request of the Montana Fish, Wildlife and Parks Department, Allendorf began studying westslope cutthroats almost as soon as he arrived in the state in 1976, initially focusing on a population the state was maintaining in one of its hatcheries. The fish were sickly, sometimes malformed, and the state wanted to know why.

Using what was considered state-of-the-art technology at the time but is hopelessly primitive today, Allendorf analyzed the genetics of the broodstock and determined that they were suffering from a phenomenon known as inbreeding depression. Brothers, sisters, cousins mating with each other generation after generation had eliminated much of the genetic variation in the population. Some beneficial genes had likely been lost, and chance had allowed other, more harmful genes to become overabundant, resulting in fish with missing fins and other abnormalities. In some cases, they may have been suffering not because one gene was worse than another, but because of the fact that diversity itself can be beneficial. A diversity of genes involved in the immune system, for example, can make an animal healthier because it provides more blueprints for more antibodies and thus makes it easier to ward off a multitude of diseases.

Based on these results, Montana officials threw out the westslope cutts they had been keeping in the hatchery. They caught more wild fish, established a new broodstock, and this time vowed to interbreed them with wild fish as often as necessary to keep them from getting too inbred.

A few years later, the agency approached Allendorf again, this time asking him to examine westslope cutts in the wild. For many decades, agency biologists had known that hybridization, usually with rainbows, occasionally with other cutthroat subspecies, was occurring in many of the waters in which these nonnatives had once been stocked. Unlike many state fish and game agencies, the Montana Department of Fish, Wildlife, and Parks has for decades taken a very progressive stance on protecting native species; it has long sought to save the westslope cutthroats. But since it is so difficult to visually identify hybrid fish, no one had a clear idea about the problem's extent.[21]

Once more the department enlisted Allendorf to examine their DNA. After several years of intensive collecting and laboratory work, he and his collaborators showed that the hybridized fish were much more widespread than anybody thought. Many of the populations that Montana officials had believed to be pure westslope cutts were in fact full of hybrids. They were, as Allendorf calls them, hybrid swarms.[22]

With these results in hand, he took the line of questioning one step further. In conjunction with several colleagues, he asked what effects hybridization might be having on the fish themselves. The experiments they conducted showed that hybrid fish grew more slowly and were dramatically less fit than their purebred cousins. Incredibly, in one study conducted on wild fish in a small tributary of the Flathead, they found that even trout that were as little as 20 percent rainbow (80 percent westslope cutthroat) were at a significant disadvantage; they had only half as many offspring as cutthroats that had not been hybridized. Allendorf argues that there are several reasons one might expect hybridization to reduce their fitness like this.[23]

For one thing, water bodies may differ in critical ways, even those separated by only a few hundred yards of dry ground. Over millennia, evolution may cause the fish to adapt to their local conditions. The need for camouflage, for example, might cause different populations to evolve unique patterns or colors that blend in with the bottom of the stream. Mix in another strain of fish, with different colorations, and their offspring may not be as well matched, reducing the fitness of the entire population. For westslope cutthroat, such local adaptations appear to have been particularly important.

Related to the idea of local adaptation are two other terms that Allendorf often bandies about: outbreeding depression and coadapted gene complexes. The idea is that natural selection will act not just on one, but on a combination of genes, so that evolution will eventually yield a set of them that all work together. During the course of an organism's life, each gene turns on and off at just the right time, and in just the right way, in an exquisitely choreographed ballet. Throw another dancer into the mix, even if his name is Baryshnikov, and the whole show will plunge into chaos if he is trying to do *The Nutcracker* while everyone else is dancing *Swan Lake*.

Using another metaphor, imagine making some hybrid words by crossing, let's say, "chair" with "table." Swap a few of the letters, which in this case correspond to genes, and you might come up with a batch of offspring like "cabir," "thalr," and "taale." None of the letters have been lost—all of those present in the parental generation are also present in the population of offspring—but the result is a hybrid swarm that is unlikely to survive very well when it's time to run the spellchecker.

Outbreeding depression, in other words, can result either from the loss of genes that are well suited for a particular habitat or from the scrambling of well-tuned gene complexes. Frequently, the effects are subtle and unpredictable. In some cases, the organism may simply grow at a slower rate or have fewer offspring. (Such a result can, ironically, compound the problem; the degraded fishing often results in even more stocking and even more hybrids.) In other cases, the disadvantages may not become apparent except under unusual and highly stressful situations, like a one-in-one-hundred-year drought. But even such rare and subtle effects can have a profound impact on a population over long time scales—potentially enough to lead to extinction.[24]

On the other hand, hybridization can sometimes be helpful. Wild populations can suffer from inbreeding depression, just like Montana's hatchery fish, especially when they are small and cut off from their neighbors, as many of the remaining westslope cutthroats are. Introducing new genes through hybridization is one way out of the rut.

Hybridization, though, especially across species and in the wild, usually ends up in the debit column of nature's ledger book. The spread of hatchery rainbow genes into wild populations, for example, threatens to homogenize the trout across wide swaths of the globe. We risk exchanging, as Allendorf and his colleague Robb Leary put it, "all of the diversity within and between many separate lineages, produced by millions of years of evolution, consisting of taxa capable of existing from the Arctic to the desert, for a single new mongrel species"—a species, they note only half in jest, that might well be called *Salmo ubiquiti*. The results could be catastrophic, should those mongrels suddenly find themselves under attack from something like disease or changing climatic conditions.[25]

Finally, hybridization can also cause another problem, perhaps the most important of all as far as many managers are concerned. Hybridization can raise difficult questions about whether a population should qualify for protection under the Endangered Species Act.

WHEN IT PASSED THE Endangered Species Act in 1973, Congress gave no clear guidelines on what should be done with hybrids. Legislators were far more concerned with the outright destruction of iconic species like the bald eagle and the blue whale, and hybridization probably wasn't even on their radar screens. Even among biologists concerned with conservation, it was a poorly understood issue.[26]

As a result, when the question first arose, the matter was referred to lawyers at the Department of Interior. In a memo published in 1981, they concluded that, as a matter of law, hybrids were not protected under the act. It was a disastrous decision, or at least poorly timed, because it was issued just as a technological revolution was making it possible to examine DNA in ways never before imagined. As biologists rushed to examine the genetics of everything they could get their hands on, it became apparent that hybridization was far more common than anyone had understood. A little coyote DNA was found in gray wolves from the northern United States and Canada, prompting calls for their delisting. And other previously protected species also found themselves in legal limbo.[27]

At the agencies responsible for administering the act, panic set in, and in 1990 they abandoned the idea that hybrids could not be protected. Agency lawyers voided their previous opinion and then washed their hands of the issue, declaring that it was "more properly a biological matter than a legal one." It was probably a smart decision. The only problem is that biologists have been unable to agree on a new set of rules, which means that it has been handled on a case-by-case basis ever since.[28]

Pity, then, U.S. Fish and Wildlife Service biologist Lynn Kaeding—the man who came into his office in Bozeman, Montana, one day in 1997 and found a petition on his desk to list the westslope cutthroat under the Endangered Species Act. Hybridization was obviously the most important factor in the species' decline across much of its range, but Kaeding had no clear guidelines from higher up about how to handle the issue. At the same time, it was a political hot potato: if his agency determined that the fish warranted listing, it would provoke screams of protest from ranchers, loggers, (your special interest group here). Powerful congressmembers would likely decry the decision and threaten to gut the act itself. The agency would be forced to devote millions of dollars to defending its decision and trying, probably fruitlessly, to turn back the tide. But if the agency decided that the species did not warrant listing, the petitioners, including a local chapter of Trout Unlimited, would almost certainly sue. And since the Endangered Species Act does not allow the agency to base its decisions on political or economic considerations, there's a good chance the petitioners would win.[29]

I don't know the inside story of just how the decision was made. It seems likely that an issue with such widespread ramifications would have reached the highest levels of the Clinton administration. Whatever the case may be, Kaeding was the primary author of the final decision declining to list the westslope cutthroat. A lawsuit predictably followed; a judge nullified the decision because it handled the hybrid issue in a way that was "arbitrary and capricious"; Kaeding went back to the drawing board, declared that, because they couldn't be distinguished

with the naked eye, any fish with more than 80 percent westslope genes would be defined as a westslope cutt; determined
that there were plenty of these fish still around; and issued another finding that listing was not warranted. And though a federal judge concurred with Kaeding, it's a saga that will likely
be ongoing for many years to come, because the fundamental
issues are not going away.[30]

Should they conclude that hybrids do not count as westslope cutthroat? In that case, the hybrids must be considered
a threat because, unchecked, they will eventually interbreed
with many of the pure populations. Should the U.S. Fish and
Wildlife Service try to poison them out of places like the North
Fork of the Flathead in a reprisal of the Green River project of
1962? Another option might be to try stocking the stream with
so many westslope cutthroats that they swamp out the rainbow
genes. But where would they get the fish? Hatchery westslopes
would have many of the same problems as hatchery rainbows.
And westslopes from different watersheds likely have different
genes—stocking them into foreign streams could result in a hybridization problem of its own.

Or should hybrids with some minimal level of rainbow
genes be counted as westslope cutthroats? There is a legal precedent for this with other species like the gray wolf and the Florida
panther. Kaeding argues that there is some biological precedent
as well: in those rare instances where their ranges overlap, the
rainbows and westslope cutts do naturally hybridize. It might
even make sense from a conservation standpoint. The remaining populations of pure westslope cutthroat are so small that
inbreeding depression is a serious threat, just as it was to the
Montana broodstock that Allendorf studied in the 1980s. The
hybrid populations contain a storehouse of westslope genes that
cannot be found anywhere else, genes that could be used to revitalize the species. But if so, what level of hybridization would
be considered acceptable? And what should be done if a population of hybrids now defined as westslope cutts threatens to
interbreed with a pure population, as they are in the North Fork

of the Flathead? Can a fish be considered a member of a species and a threat to the species at the same time?[31]

What about the resources that society would have to devote to saving these fish if they were listed? When I spoke to him, even Allendorf acknowledged that the westslope cutthroat probably couldn't be recovered. He and others believe the money and the political capital could probably be better spent elsewhere—perhaps protecting another declining species of fish like Montana's bull trout. But that's not the point. As Allendorf told me, "I think the Fish and Wildlife Service realized that if westslope were listed, it would be a huge nightmare. And so I think what they began to do—they knew what the answer was that they wanted—and they just distorted the evidence that was out there to support what they wanted to do."[32]

What is troubling, in other words, is not that the Fish and Wildlife Service prioritized as it did, but that it seemed to distort the science in order to do so. It sets a precedent that, if accepted as the norm, could pose problems for other species. But what are the alternatives? Should the Endangered Species Act be changed to allow economic and political considerations to play a determining role? Doing so would surely open a Pandora's box and probably have the effect of gutting the act. But not doing so may force the agency to clothe political decisions with scientific justification and result in fruitless arguments and never-ending lawsuits. Resolution is hard to come by when the debate is on one level and the true source of conflict on another.

A MONTH OR TWO AFTER meeting Allendorf, I decided I needed to do a little more research on the matter. My wife grew suspicious when she saw me packing my fishing gear, but after some explanation (and only a very small amount of obfuscation), she let me go. I kissed her and our three young children goodbye and hit the road for the North Fork of the Flathead.

When I saw a fly shop about ten miles from my destination, I stopped in to get some advice and pick up some flies.

The owner handed me an enormous hairy orange thing with a hook poking out the back that he called a "stimulator." What was it supposed to look like? "A little bit of everything." I was skeptical; most of my fishing of late has been with small nymphs (flies that bounce along underwater). It's not quite as enjoyable as catching fish with dry flies, but in many cases it's the only way to hook anything. The owner of the shop assured me that wasn't necessary. We only fish with dry flies around here, he informed me. Who's we? "Everybody."

And so, on a morning in mid-August, I found myself hurling my newly purchased orange fluff into the North Fork of the Flathead. It is a gorgeous river. Cradled by a broad valley, the main stem of the river slides cold and clear over broad riffles and pools up to a dozen feet deep. As you progress downriver, fast headwater streams occasionally tumble into the valley from the massive gray peaks of Glacier National Park on the east and the only slightly more gentle Whitefish Range on the west. In the summer, late in the evening, when the sun finally sets, a beautiful golden light illuminates the cobbles and willows along the river's banks. Home to grizzly bears, mountain lions, wolves, and very few people, the valley remains one of the wildest and surely most scenic places in the lower forty-eight.

Not that there was much time to enjoy the scenery. The stimulator produced almost instantaneous results. According to fisheries biologists, cutthroats are some of the most easily caught trout in the world, and these westslopes seemed eager to prove them right.[33]

On my way down to the river, I had passed by a sign produced by the Montana Department of Fish, Wildlife, and Parks. The sign instructed me to release native fish like the westslope cutthroat, but urged me to keep the nonnatives I caught. It helpfully provided a picture of a native westslope cutthroat and a nonnative rainbow, along with a couple of other fish, the native bull trout and the nonnative lake trout. A red or orange slash beneath their jaw characterized the westslopes, according to the sign, as did the pattern of black spots that grew more dense on the tail and sparse to nonexistent on the belly. Rainbows, on the other hand, were characterized by a relatively even scattering of spots, even on their belly, and by a red or pink side band.

If only it were so clear-cut. The first three fish I caught seemed unambiguous. They flaunted their bright orange slash, and a cloud of black spots arced up over their blushing pink belly and dissipated toward the head. The difficulty lay with the fourth one. Orange slash, yes. But the black spots progressed from tail to head without even a pretense of avoiding the belly. And was that a red stripe along the lateral line? What was I to do with this fish?

Shortly before departing for my trip to Montana, I had been procrastinating, scrolling through the science blogs. I had come across an article about an experiment in which a team of neuroscientists set up brain-imaging equipment, showed volunteers a sequence of letters on an otherwise blank screen, and asked them to choose one of two buttons whenever they desired. It didn't really matter which button they chose, only that they made a decision, and when they had done so, pushed. Based on the brain activity they were observing, the experimenters found they were able to predict a person's choice well before the subject was aware of making a decision—sometimes by as much as ten seconds. In other words, the experiment showed that even though our conscious brain may think it's in charge and decisions are the product of rational thought, it is instead the subconscious that often occupies the driver's seat.[34]

What does that have to do with westslope cutthroats? Much of the literature on westslope cutthroat conservation, and many of the discussions I've had on the dilemmas posed by these fish, centered on science and the law. Will hybridization yield fish that are less fit than their purebred counterparts? What does the Endangered Species Act require?

These are important questions, no doubt, and they deserve discussion. Ultimately, though, it seems that something deeper, something unsaid, often drives the debate over these fish. One such undercurrent is the philosophy that, in my mind, will forever be associated with a picture of the seemingly always smiling face of a woman named Helen Chenoweth. In 1995, when I was a reporter in Idaho, Chenoweth was serving her first term as one of the state's congressional representatives. An unapologetic member of the God-gave-man-dominion-over-the-earth school of thought, Chenoweth always seemed to derive great pleasure from tweaking the environmentalists. During her campaign, she protested the Endangered Species Act by holding what she called "endangered-salmon bakes" for her supporters. And (at least partly, I suspect, because she knew it would drive her opponents wild) she once declared that salmon obviously couldn't be endangered since it was so easy to buy the canned variety in the local supermarket.[35]

Another current is the idea that science and technology have led to "the end of nature." It's an age-old idea, but this distrust of science and technology seems to have become an especially potent force in the American subconscious since the early 1960s, when "better living through chemistry" gave way to *Silent Spring.* It is a philosophy that is perhaps most visible today in the often vociferous opposition to genetically modified organisms, and which is easily transferred to hybridized westslopes that are, in their own way, GMOs too.

These are unquestionably oversimplifications, ideas that for other people might be divided and articulated in other ways. Few people are easily pigeonholed into one camp or the other; most would probably acknowledge that multiple philosophies play a role in their thinking. The point is that value judgments based on beliefs that exist well beyond any sort of logical discourse often form the basis for people's decisions.

Even in the remote North Fork, I was under no illusion that I had somehow escaped civilization. Helen Chenoweth is always on my shoulder, whispering in my ear that humans have always been an integral part of the natural world, that trying to separate the two creates a false dichotomy. But at the same time, I find it profoundly unsettling to think that humans can now manipulate DNA at will, unleashing GMOs the planet has never before seen. And it made me just as uneasy to be holding a fish that was not only a sort of GMO itself, but was also a vector of the rainbow trout genes that were transforming all the natives into something similar.

I tried to decide what to do with the being that was gasping and occasionally struggling in my hand. Then I realized my subconscious had already made the decision for me, probably many seconds ago. I gently released the hook, placed the fish back in the water, and watched it swim back into the depths.

It Doesn't Do
Any Good

John Muir liked to call California's Sierra Nevada "the Range of Light," and once you've experienced the luminous granite, sparkling waters, and brilliant sunlight of the high country, it's hard to think of those mountains in any other way. More than seven thousand lakes punctuate the range, providing welcome relief for hikers and a powerful draw for anglers from the valleys below.[1]

In the fall of 2006, I visited one of those lakes with some seasonal employees from the Forest Service and the California Department of Fish and Game. Covering about twenty acres in the middle of a high granite basin not too far from Yosemite National Park, the water was so clear that even in the middle you could see the bottom, fifty feet below. It was a stunning place. The name, though, has to remain confidential.

The "seasonals," as they are known, were on a mission: to eliminate the rainbow trout that somebody first introduced to the lake many decades ago. They had been hiking to the lake

In a dramatic turnaround, the California Department of Fish and Game has
begun using gillnets to remove rainbow trout and other fish from some of the
lakes in the Sierra Nevada that it had previously stocked. Similar operations are
ongoing throughout the country and the world.

every other week since the ice melted to tend to dozens of gillnets. Their employers fear that if the location of the lake is re-
vealed, someone will try to sabotage their efforts.

The gillnets were first stretched from the shore out into the
middle of the lake a year earlier. Made of fine monofilament,
they are almost invisible underwater. Fish that inadvertently
poke their heads through one of the holes are snared—unable
to proceed forward because of their fins and unable to back out
because their gills act like the barbs on a hook. It helps to clean
the dead fish, algae, and other debris out of the nets and sew up
the holes every so often, as the seasonals were there to do, but
gillnets are potent fish killers even when unattended.

The nets had hung under the ice like curtains all winter,
and many trapped fish probably decomposed and drifted away
without anyone being the wiser. During my visit, we found

only a few four-to-five-inch rainbow trout in the nets, in various stages of decay. This, I was told, was typical of recent visits, suggesting that not many fish remained in the lake. After several years of such labor, California Department of Fish and Game higher-ups hope to have removed every last one of the rainbow trout. It's an ironic situation, as I'm sure everyone in the agency knows. Because it was the California Department of Fish and Game that had been stocking these fish into the lake every other year until 2001. I'm getting ahead of the story, though.

THE LAKES OF THE high Sierra were formed by glaciers that rearranged the topography as they waxed and waned. The ice sheets scoured holes in some places and left huge piles of rock in others, so that when the glaciers retreated for the last time about ten millennia ago, thousands of lakes formed in their wake. These lakes must have been oases amid the newly uncovered rock, but waterfalls separated many of them from the rivers below. Only those forms of life that could travel by land or by air could colonize them, including amphibians, insects, and other invertebrates. In the absence of fish, fierce predators and competitors that often structure the communities in which they live, distinct ecosystems took shape that were relatively undisturbed for thousands of years.

To most of the Americans who first visited the area in the late nineteenth and early twentieth centuries, however, the lack of fish meant the high mountain lakes of the Sierra Nevada were being wasted. And so fingerling trout were hauled up to many of the lakes by miners, explorers, recreational groups like the Sierra Club, and ultimately by the California Department of Fish and Game. The fish were carried on mules and in backpacks, in milk cans, coffee cans, and any other kind of container that could hold water.[2]

Unfortunately, for most lakes, it is impossible to say when they were first stocked, or with what species. The stocking efforts, even by California Fish and Game, were sporadic and unorganized, and few records were kept. Brook trout, first sent to California in exchange for rainbows in the nineteenth century, are the most common trout in the high mountain lakes today, but golden and rainbow trout are also prevalent. It appears, however, that no matter how much stocking was done, it was not enough to satisfy some visitors. "There are more lakes and streams in the High Sierra of eastern Fresno County than in any other area of the same size in the State of California," one Fish and Game employee noted in 1929. "The fact that most of these waters have continued barren of fish life has caused comment by the many tourists and campers who annually visit this wonderland, and has brought criticism of the Fish and Game Commission for its neglect to develop this important region."[3]

The California Department of Fish and Game apparently set to work to rectify this situation, and one year later, the same man reported, "It is to be hoped that this work will progress so when these now inaccessible areas are open to automobile and airplane travel, vacationists will find an abundant supply of fish awaiting their arrival."[4]

Little could he have known how much progress the department would soon make. With their newly developed aerial fish-stocking techniques, the California Department of Fish and Game carpet bombed the lakes of the high Sierra in the years that followed World War II—a pattern that was repeated throughout the western United States. According to one estimate, probably only 5 percent of the thousands of high mountain lakes in the West contained fish a hundred years ago. Today, only about 40 percent lack fish, and most of these remain in that condition because they are too shallow to support fish on a year-round basis. About nineteen out of twenty of the big and deep lakes have fish.[5]

There were surprisingly few objections to the stocking programs. Environmental groups like the Sierra Club seemed to take their cue from Muir, who encouraged fish stocking in the Sierra Nevada because he knew it would create a constituency for the region. "In this case she [Nature] waited for the agency of

After years of stocking the lakes of the Sierra Nevada using pack animals, the California Department of Fish and Game began using airplanes in the 1940s. Many of the lakes previously had no fish.

Courtesy of Carrol Faist, California Department of Fish and Game

man, and now many of these hitherto fishless lakes and streams are full of fine trout," Muir wrote in 1901. "Soon, it would seem, all the streams of the range will be enriched by these lively fish." Fish stocking, he wrote, "will become the means of drawing thousands of visitors into the mountains. . . . Trout-fishing regarded as bait for catching men, for the saving of both body and soul, is important, and deserves all the expense and care bestowed on it."[6]

There was some concern from the scientific community about the impacts of such practices. The naturalist Joseph Grinnell noted as early as 1924 that the stocking of fish into the high mountain lakes was affecting amphibians and other inhabitants. For the most part, though, there was very little scientific objection to the introduction of fish into high mountain lakes from academia or elsewhere. It seems to have been a case of out of sight, out of mind. Few scientists studied the high mountain lakes—not only are they remote, but the life that exists in them is underwater and thus invisible from the shore.[7]

And among the fisheries managers from the California Fish and Game Department, few questioned the central assumption on which the stocking program was predicated—that the high mountain lakes were simply too harsh to support a worthwhile fishery without a constant infusion of hatchery fish. Indeed, there was almost no research into the effects or the effectiveness of the stocking program at all.[8]

Why did the agency conduct so little research on a program that was so expensive and widespread? It was, according to one former employee of the agency, simple bureaucratic intransigence. And indeed, with millions of dollars invested in airplanes and hatcheries, with employees who depended on the continuation of the aerial stocking program, and with the backing of political heavyweights and a public to which it had long promoted the virtues of the program, it is easy to see why the California Department of Fish and Game would have had a hard time entertaining questions about the aerial stocking program. And so, with very little research on its effectiveness, aerial stocking of the high mountain lakes in the Sierra Nevada continued on an intensive and ongoing basis until, in 2001, it came to a near-total stop.[9]

THE U.S. FOREST SERVICE, which administers much of the land in the Sierra Nevada, has long deferred to the states when it comes to fish and wildlife management. Even in designated wilderness, which was defined in the Wilderness Act of 1964 as "an area of undeveloped Federal land retaining its primeval character and influence . . . [that] generally appears to have been affected primarily by the forces of nature, with the imprint of man's work substantially unnoticeable," the Forest Service has typically continued to allow aerial stocking of any lakes that had been so stocked prior to wilderness designation.[10]

In 1995, however, Forest Service officials were rewriting their management plans for some of the wilderness areas in the Sierras and they wanted to know more about the status of the alpine lakes and their inhabitants. They were particularly concerned about the mountain yellow-legged frog (*Rana muscosa*). A medium-sized brown frog endemic to the mountains of California and Nevada, the mountain yellow-legged frog was once the most common amphibian in the Sierra Nevada. The frogs were so abundant that members of a survey team in 1915 reported that they could not help stepping on them as they navigated their way around some lakes.[11]

By the end of the twentieth century, however, the frogs had disappeared from much of their former range. One researcher estimated that they were present in less than 15 percent of the lakes in which they were reported in 1915. Work by some researchers suggested that fish stocking might be to blame, while others thought it might be due to pollution or the drift of pesticides from the Central Valley. Because no wide-scale, systematic surveys had been done, however, it was difficult to assess the size or extent of the decline or when or why it had occurred. Consequently, the Forest Service decided to hire a scientist by the name of Roland Knapp to survey the alpine lakes in some of the wilderness areas it administered.[12]

For his Ph.D., Knapp had studied the behavioral ecology of tropical fish—how do females choose their mates, does it make any difference in the survival of their offspring, that sort of thing. After he got his degree, though, he decided he wanted his research to have a more practical application, and he took a job at the Sierra Nevada Aquatic Research Laboratory, a.k.a. SNARL—a research facility on the eastern side of the mountains that was started in the early part of the century as a federal fishery research center and had eventually become part of the University of California.[13]

Knapp agreed to survey the lakes in places like the John Muir Wilderness and the Ansel Adams Wilderness. When he started, he expected to be looking at five hundred to a thousand lakes over a couple of years. Instead, ten years later, Knapp and his team had surveyed every water body in the Sierra Nevada big enough to show up as a blue dot on a 7.5-minute topographic map between Mount Whitney and Yosemite. That's about seven thousand lakes and ponds in both the national parks and the national forests.

These were not cursory surveys. They included physical measurements of size and depth, description of the morphology and substrate, counts of amphibians and reptiles, presence and size of fish barriers and spawning areas, and samples of the invertebrate life of the lake or pond. If the lake was deep, the surveyors also went snorkeling and laid out gillnets to establish not just the presence or absence of fish, but the species composition and population structure as well.

At the same time that they were conducting the surveys, Knapp and a colleague received a grant from the National Science Foundation to do some experimental work on the effects of fish in the high mountain lakes. In a remote basin in the Sierra they used gillnets to remove the fish from five of the eight fish-containing lakes and then watched to see what would happen. The results from both the surveys and the experiments left little room for doubt: fish and frogs don't mix. Frogs were seldom found in lakes containing fish, and when the fish were removed, the frogs often made a dramatic comeback.

When I visited Knapp in the Sierras one summer, he took me to visit one of the lakes from which the fish had been removed a few years earlier. On my hike in, I had seen no amphibians in any of the many bodies of water I passed. But around that lake, the frogs were thick, hopping from underfoot into the water at the last possible moment. The tadpoles were clustered in warm water pockets along the edge, and there were so many in some places that they completely obscured the mud below. According to Knapp, about two thousand adult frogs lived in the lake, a dramatic increase over the twenty or so that were there before the fish were removed.

Exactly why the fish cause the frogs to disappear is not entirely clear. They might spread disease or they might outcompete the tadpoles for food. Most likely, though, the mechanism by which the fish eliminate the frogs is a little more direct. They eat them. With the frogs having evolved over the last ten thousand years in blissful ignorance of fish, it is probably no surprise that they are so vulnerable to these fearsome omnivores. In addition, the mountain yellow-legged frog is a particularly aquatic species of amphibian, making them all the more vulnerable to aquatic predators. The adults are seldom found more than a couple of hops from a lake or pond during the summer, and they spend their winters in the water under the ice. More important, the tadpoles spend at least two years in the lakes before they metamorphose into frogs. This means they can only live in deep lakes that do not freeze all the way through during the tough Sierra winters. As fish have similar criteria, these are the same lakes that were most commonly stocked.[14]

Even lakes that were never stocked felt the impact. Amphibian populations naturally fluctuate widely in size, and occasionally the number of frogs in a lake is bound to go to zero due to chance alone. Under normal circumstances, when this happens, the lake will likely be recolonized by frogs from a neighboring population. If, on the other hand, the frogs are permanently eliminated in all of the neighboring lakes due to the presence of fish, then the frogs in a fishless lake will also eventually disappear when bad luck wipes out the existing population and the lake cannot be recolonized. Another group of researchers in Sequoia and Kings Canyon National Parks estimated that the introduction of fish to that region of the Sierra had cut connectivity between lakes and thus the number of potential colonists to a tenth of its former level.[15]

It is not just the frogs that have been affected, either. According to Knapp, if you decided to put on a snorkel and take a swim in some of these frigid lakes, you would find that the difference between those with fish and those without is stark. At another lake in the Humphrey's Basin from which the fish had been removed, Knapp showed me a larval mayfly, *Callibaetis ferrugineus*. This aquatic invertebrate was completely absent when the lake contained fish. Now there are between one hundred and five hundred in every square yard, and lots of other species as well. According to Knapp, the number of species in the lake has increased tenfold and the total biomass of the invertebrates has increased by a factor of one hundred.

And since the mayflies and other invertebrates are often eaten by other animals, such as birds and bats, after they emerge from the water and take flight, the presence of fish in a lake may have impacts that spread into the terrestrial realm as well. As but one example, Knapp pointed specifically to the gray-crowned rosy finches that liven the Sierra Nevada during the summer months. These birds are often hatching their eggs around the time of some of the biggest mayfly hatches and can be seen during this critical time collecting these insects from fishless lakes to bring back to their young. Hundreds of thousands to millions of these mayflies can emerge from a fishless lake over a two-week period in the spring while none emerge from those lakes containing fish. A protein and energy subsidy like this could make a big difference to these birds.[16]

More importantly, at least in some quarters, Knapp's data suggested that the fish populations in many of the lakes were probably self-sustaining. Because they are cold blooded, fish have a much lower metabolism than warm-blooded animals. That means they can survive and even thrive in lakes that are covered with ice for a majority of the year and have a limited food supply. Further, with only a few square yards of well oxygenated sand and gravel in which to spawn, a few fish can produce more offspring than could ever fit into the cargo hold of a plane. Once the fish are introduced to a lake, therefore, stocking may be unnecessary for the maintenance of good fishing. It may, in fact, even make the fishing worse. For if there are too many fish in a lake, none of them will have enough food, resulting in a lake full of stunted fish that few anglers would waste their time hiking for miles to catch.[17]

In 1995, when Knapp and his team were surveying the lakes, the California Department of Fish and Game agreed not to stock those lakes that were to be surveyed. Nevertheless, the surveyors found fry in many of the lakes. Since they did not come from a plane they must have come from natural reproduction by fish that had overwintered in the lake. Based on these preliminary data, the California Department of Fish and Game agreed to stop stocking sixty lakes for a period of five years to determine whether the fish population could, in fact, sustain itself. The work showed conclusively that about 70 percent of the lakes that the department had been stocking for fifty years or more held self-sustaining trout populations. In other words, the agency had at best been throwing money away—up to a million dollars a year according to Knapp—and at worst spending that much to make fishing worse by creating lakes full of stunted fish.

How was this analysis received? "When I first proposed this work to Fish and Game in 1995," Knapp recalled, "they were sort of indifferent to the work. They were basically saying there are no effects in the backcountry of these fish, so why do you want to do this work? Yeah, you can have permits, but it really doesn't have any bearing on what we are doing, because there are no effects."

The tone changed the next year. "As soon as those results began to come in, their posture became much more defensive," Knapp claimed. The reaction, at least among a certain portion of the department, was something along the lines of, "'How can this be? How can you be seeing all these effects? Something must be wrong . . . who's this crazy scientist that's coming in, telling us that these fish are doing bad things to the backcountry?'"

That was the low point of his relationship with the department, Knapp said, but nevertheless there was some cautious interest in his work. For one thing, department officials were worried about the possible listing of the mountain yellow-legged frog under the Endangered Species Act and the consequent loss of control to the federal government. Knapp's experiments showed that fish removal could lead to a recovery of this and other species. If the state government could show that it was helping the frogs to recover, or at least trying to, it might avoid a listing and retain control of the stocking program.[18]

And if the link between fish stocking and the decline of the mountain yellow-legged frog was not enough to convince the department to curtail the stocking program, the economic argument was. Department officials drew up some internal regulations that put a stop to stocking in areas where there was no "valid or reasonable fisheries management justification," their way of saying lakes with self-sustaining fisheries. By 2001, a department that used to stock more than one thousand lakes in the Sierra Nevada every year was stocking fewer than twenty. Money that had been spent on the stocking program instead went toward surveying the lakes that Knapp and his crew had not yet done. And perhaps most significantly, the California Department of Fish and Game began drawing up plans to eradicate the fish from some of the lakes that it had been stocking only a few years earlier.[19]

PHIL PISTER HAS SPENT the vast majority of his seventy-seven years in Bishop, California, a small town on the eastern side of the Sierra, about ninety miles south of Bridgeport. For much of that time, he was an employee of the California Department of Fish and Game, charged with managing the fisheries in the region. When he began working for the department in 1952, Pister's job was to determine how many fish should be planted in each body of water. He would make a list and hand it to the people at the hatcheries, who would, in turn, load them up in a truck or an airplane and drop them in. "The thinking back in the '60s, the early '60s, by fish and wildlife people, particularly in the state agencies, was if you can't catch it and eat it, the hell with it," Pister recalled. "Back then the entire effort in virtually all fish and wildlife agencies in the West, especially in California, was to provide sportfishing." At the time, Pister scoffed at the idea that planting fish in the previously fishless high Sierra lakes could do any harm to the ecosystem.[20]

With the exception of a few golden trout, most of the fish Pister and his partners planted in roadside as well as backcountry waters were rainbows of various strains. The department built several large hatcheries to raise these fish, and the eastern Sierra rapidly became one of the most famous fishing spots in the state, especially among the anglers from Los Angeles and other parts of southern California.

In 1965, however, Pister received a phone call from Robert Miller and Carl Hubbs, the same two ichthyologists who had fought the Green River rehabilitation project with such vehemence. Miller had studied some of the native fish of the region for his doctoral dissertation, and he and Hubbs were interested in returning to the area to look for one of these species, the Owens pupfish. Many of the region's wetlands had dried up, thanks to the Los Angeles Aqueduct, and introduced fish also spelled trouble, but the pair had some hope that the fish might still exist in the warm, shallow waters of a marsh near Bishop known as Fish Slough.[21]

Pister agreed to take a day off from his usual duties and accompany them, a fateful decision that would completely change
the direction of his career. For when they found some of the
pupfish, one of the last populations, a new sense of purpose
gripped Pister. And from that day on, Pister began to devote
much of his time to the conservation of these and other desert
fish—fish that had no commercial or recreational utility.

Pister was one of the first officials in any state fish and game
agency to take an active interest in nongame fish. And as an
employee of the agency, Pister also tried to force the California
Department of Fish and Game to reconsider its role. It was not
always an easy sell. Outside of a few ichthyologists in academia
like Miller and Hubbs, there were few people who were concerned with nongame fishes at that time. Frequently, even the
big environmental groups like the Audubon Society, Defenders
of Wildlife, the Sierra Club, and The Nature Conservancy failed
to recognize the value of such fish.[22]

Nevertheless, he did have some successes. In 1968, Pister,
Hubbs, and Miller convinced the California Fish and Game Commission and the Los Angeles Department of Water and Power
(the landowner) to establish the Owens Valley Native Fish Sanctuary to preserve the four native fish of the area. In 1971, Pister
and Miller wrote a paper about the sanctuary in the *Transactions
of the American Fisheries Society*. This publication has been the
preeminent fisheries management journal in the United States
since it was first issued in 1872, yet their paper was the first it
ever printed about nongame or noncommercial species.[23]

Pister's success in the Owens Valley gained him some noto-
riety in certain circles, and before long he and others had formed
a group known as the Desert Fishes Council to try to preserve
other native fishes. To the dismay of his superiors, Pister became
an early advocate of federal intervention on behalf of threatened
species, because he believed this was the only way to overcome
the complicated jurisdictional issues that made conservation
so difficult. He wrote letters and made phone calls to highly
placed officials in the Interior Department, informing them of

the threats faced by some of the native fish of the region and urging them to take action. He also flew to Washington to work on drafts of the Endangered Species Act before it became law in 1973.[24]

Because he was three hundred miles from his boss in Sacramento, Pister was able to get away with some of these controversial activities. But he also ran into some trouble and difficulties. On one occasion, a letter he had written to an assistant secretary of the interior fell into the hands of the director of the California Department of Fish and Game. The letter supported federal intervention on behalf of threatened fish and was none too popular in Sacramento. Shortly after finding out about it, the director and other higher-ups in the department summoned Pister and told him in no uncertain terms to cease and desist. "I was in the army right after World War II, long enough to know you never got off of KP by fighting with the general," Pister recalled. "So I said OK. And I went underground. I continued doing the same thing, I just made less of a point of talking about it or writing about it."

On another occasion five or ten years later, Pister decided it would be worthwhile to evaluate the impact of aerial fish planting on high mountain lakes. He put a proposal together and sent it to departmental headquarters in Sacramento. Sometime thereafter he got a call from the chief, who said something like, "I agree you ought to do this. But the director does not want to hear that his personal airliner is going to be grounded because we can't plant fish with it anymore. The best thing to do is just to forget this." The fish-stocking plane, Pister recalled, was a gorgeous aircraft—a Beechcraft King Air—and it was a popular means of transport among state officials, including the governor. Pister decided to drop this line of research.

Reflecting on his career in 1991, Pister wrote that one could not help but wonder "why professional biologists employed by western fish and wildlife management agencies lacked the motivation to inventory their native faunas and devise programs to ensure their perpetuation. Having spent most of the past two decades pondering this question, and based on my own experiences as a state fish and game agency biologist, I would lay most of the blame at the feet of a bureaucracy rooted in tradition, an almost universal program direction and professional ethic built around sport and commercial fishing, and university curricula devoid of courses in environmental ethics and ecological and evolutionary principles."[25]

ULTIMATELY, OF COURSE, the bureaucracy has changed. The gillnets and about twenty lakes in the eastern Sierra that are newly devoid of fish can bear witness to that. Just the existence of an eradication program represents a monumental shift in direction for an agency that not so long ago measured its success by how many lakes it had stocked. Even more broadly, the California Department of Fish and Game now employs dozens of nongame biologists, and allocates by far the biggest portion of its funds to biodiversity conservation.[26]

It's not just California, either. Most other states have undergone similar shifts. Fish removal and restoration projects are planned or have already occurred in states from Idaho to North Carolina. Federal agencies like the National Park Service and the U.S. Fish and Wildlife Service, which spent much of the first half of the twentieth century stocking rainbows into waters that had never seen their kind before, are today taking a leading role in removing them.[27]

Partly, no doubt, this shift occurred only because of individuals like Pister, Miller, and Hubbs. Partly, the shifting climate within the agencies has been driven by the threat of federal intervention. State officials recognize that if species like the mountain yellow-legged frog become listed under the Endangered Species Act, they will lose control over the natural resources in their state. And such listings can impact not just the fishing, but other industries as well, from agriculture to computer chip production. Fish and game commissions in California and elsewhere have thus been forced by elected officials to take action to protect those species at risk of ending up on the list.

Finally, though, society as a whole has undergone a dramatic shift. Nongame species have developed a large constituency. Groups dedicated to the preservation of native wildlife have boomed in the past several decades. Bird watching and other forms of outdoor recreation that do not depend on hooks or bullets have grown. And, at the same time, the popularity of sportfishing and hunting has leveled off. The total number of fishing licenses sold in California peaked in 1981 and has never been as high since.[28]

The changing role they are forced to play has led to controversy and division in the state agencies. Although employees of the California Department of Fish and Game were hesitant to acknowledge it, there seems no question that the fish removal program has generated conflict within that agency. "This was hard," Pister told me, "particularly for the pilots. You know these guys, they work their heads off and make sure that they've planted all the lakes just right—did everything according to protocol. And then all of a sudden, you say, 'What you guys have been doing, it doesn't do any good.'"

The shift in emphasis has not been popular with the traditional constituents of the fish and wildlife agencies. For one thing, hunters and anglers argue that they are supplying most of the funding for the state fish and game agencies through license fees and taxes on fishing and hunting equipment. It is not only unfair to make the sportsmen subsidize the preservation of nongame species, but it also means there is less money available to the departments to manage these activities. And if that wasn't enough, the anglers are now being forced to fund the eradication of the very fish they pursue.

And if the anglers are unhappy, many of the business owners who depend on them are nearly apoplectic. Take, for example, Bridgeport, California. Like many towns on the eastern side of the Sierra, Bridgeport depends heavily on recreation, especially on the millions of anglers from central and southern California who come to fish the nearby reservoirs and the lakes and streams of the Sierra every year.

I visited Bridgeport after one of my trips into the Sierras, and when I casually described the gillnets to the owner of the hotel in which I stayed, I got an earful. My notes from the conversation are full of scrawled quotations: "idiots," "stupid blunders," "stepping on their own toes," and other unreadable scribbles. And as I walked around town the next day, talking to the owner of the fly shop, my server at the diner, and others, it quickly became apparent that his sentiments were widely shared. From what I could tell, the local business owners and residents were upset: (1) because the department was removing fish; (2) because they heard about it from an outsider like me; and (3) because they feared that the public would think twice about going on an expensive fishing trip in a region where fish were being actively eradicated. "You get one bad story about fishing and people will go somewhere else."

How does the Department of Fish and Game respond? Curtis Milliron supervises the fish removal program in the eastern Sierras for the agency. When I called him and began asking about the "eradication program," he calmly and professionally sought to set me straight. Fish removals, he told me, were but one small part of a new way of managing the region based on aquatic biodiversity management plans—blueprints for each watershed that seek to balance recreational fisheries with the needs of all the native species. Sure, the mountain yellow-legged frogs and Knapp's work kick-started the program, but the department was not just running around killing fish in some haphazard attempt to save the latest poster species for the Endangered Species Act. These were carefully thought out plans based on mountains of data about the biological and recreational characteristics of every watershed.[29]

The department hadn't publicized the specific lakes in which the removals were occurring because it feared sabotage. In fact, some of the gillnets had been destroyed, apparently by someone who did not want the fish removed. Even more, Milliron feared the bucket brigade. If somebody wanted to, it would be a simple matter to catch a few fish from another lake, put them in a bucket, and carry them over to one of the lakes in which fish had recently been eradicated. Natural reproduction would quickly destroy all of the department's efforts. It's not an implausible scenario. Such illegal fish introductions have become increasingly problematic around the United States. And, as Milliron points out, that is exactly how entire basins in the high Sierra Nevada became populated with fish only a hundred years ago.[30]

Nobody need fear for the quality of the fishing in the eastern Sierra, either, Milliron claimed. For one thing, eliminating fish from the high lakes would be physically and politically impossible. At an absolute maximum, only about 150 of the 700 lakes in Milliron's purview that currently contain fish will be targeted. And even that is unlikely anytime in the near future without an unlimited supply of money and personnel.

In truth, it's the same story throughout the country. Fish removal projects are occurring everywhere, often with lots of publicity. But if you compare the waters from which they are being removed with the waters where they have been introduced and established self-sustaining populations, it's just a drop in the bucket. Rainbow trout, it seems, are here to stay.

The Last Generation
of Troutfishers

More than a century ago, a famous Colorado preacher and politician named Myron Reed issued a wistful lament for his favorite pastime. "This is the last generation of troutfishers. The children will not be able to find any," the reverend declared. "Not that trout will cease to be. They will be hatched by machinery and raised in ponds, and fattened on chopped liver, and grow flabby and lose their spots. The trout of the restaurant will not cease to be; but he is no more like the trout of the wild river than the fat and songless reed-bird is like the bobolink. Gross feeding and easy pond-life enervate and deprave him. The trout that the children will know only by legend is the gold-sprinkled living arrow of the white water; able to zig-zag up the cataract; able to loiter in the rapids; whose dainty meat is the glancing butterfly."[1]

I have often thought of Mr. Reed while researching this book, and wondered what his reaction would have been to the hatchery rainbows of today. What would he have said about the

Adam Konrad caught this world-record forty-three-pound rainbow trout in a Saskatchewan lake. His prize probably had been manipulated to contain an extra set of chromosomes—a feature that makes such fish grow much faster and larger than normal—and then escaped from a nearby aquaculture facility.

Courtesy of Otto and Adam Konrad

scientists from Missouri who have been feeding their rainbow trout creatine—the same muscle enhancer that baseball slugger Mark McGwire and other athletes use to achieve such incredible bulk and endurance—because they believe "there would be a lot of marketability for harder fighting fish"?[2]

How would he have responded to the pictures that circulated on the Internet not so long ago of a record rainbow trout caught in a lake in Saskatchewan? The fish, a forty-three-pound escapee from a trout farm with an extra set of chromosomes to make it grow faster and bigger, is so grotesquely obese, it's hard to imagine how it could swim.[3]

And what choice words would he have penned if he had seen the *Wall Street Journal* on August 15, 1996? At the top of the front page was an article describing an Idaho Department of Fish and Game attempt to teach its hatchery rainbows how to eat worms. The fish, it seems, had grown so accustomed to synthetic food pellets that they were not interested in the bait

anglers were throwing at them, forcing the department to edu-
cate them about their purpose in life before releasing them into
the wild.[4]

On the other hand, hatchery rainbows and stocking programs have never faced greater scrutiny than they do today. In
Reed's era, eager anglers and public officials sought to spread
rainbows into as many waters as possible, "with the hope that
the conditions would be favorable and that they would strive
with our brook trout for the mastery and that the result would
be the survival of the fittest." Little changed for almost a century, except that the discovery of rotenone and other fish poi-
sons meant fisheries managers no longer had to rely on natural
selection. They could engineer any fishery they desired.[5]

Today, it's illegal in most places to stock any fish without
a permit from the state fish and game agency, not an easy thing
to get. And even as they continue to stock rainbows, the state
and federal governments now spend millions of dollars every
year to recover natives, sometimes even poisoning the rainbows
to do so. In response to increasing demand and the pressure of
groups like Trout Unlimited, agencies are setting aside increasing numbers of streams as catch-and-release wild trout waters—
no stocking allowed. And the U.S. Fish and Wildlife Service,
descendant of Baird's U.S. Fish Commission and long the largest
grower of rainbow trout in the world, now goes out of its way to
distance itself from these fish, declaring that its primary goal in
fisheries management is to help recover and sustain threatened,
endangered, and native fish.[6]

WHEN I FIRST CONCEIVED this book, I must admit, I
wasn't fishing much. I had largely stopped sometime in the
1990s. It wasn't really a conscious decision to quit. I was living
in Colorado, which stocks more rainbows per capita than almost any other state in the country, and without really realizing
it was happening, I got bored. Going out to catch another ten-
inch stocked rainbow lost its appeal. I stopped making time for

fishing. And once I started working on my Ph.D. in ecology and evolutionary biology, I got my fix chasing frogs and tadpoles.

As I was writing this book, though, I started fishing again. I wasn't bored anymore. Hold a rainbow in your hand, and you are holding a savior of democracy (à la George Perkins Marsh), and likely a partial descendant of one of the fish Livingston Stone captured all those years ago on the banks of the McCloud. Others cleverer than I have asked whether humans were domesticated by animals like dogs and sheep, instead of the other way around. Likewise, look that fish in the eye, imagine all the effort that humans have put into helping the species achieve a nearly global conquest, and ask yourself which one of you is subordinate in the relationship.

That said, I still prefer catching not just wild fish, but natives. I applaud the groups and agencies trying to restore native fish and ecosystems. I think there is a case to be made for such programs, not just from an ecological or conservation standpoint, but from an economic perspective as well. People can catch hatchery rainbows anywhere. But there is only one place in the world where they can catch a greenback cutthroat in its native element, and someday, such unique opportunities surely will prove to be a much greater draw for anglers, and thus a more powerful engine for local economies.

I do believe, though, that those who promote the conservation and restoration of native species should do so with a good understanding of history and a concomitant sense of humility. People have been a part of this world for a long time. There's no going back to the way it was, even if it were possible to define it. Reading through the letters and public pronouncements of the men who were most responsible for spreading nonnative species like rainbow trout throughout the world in the nineteenth century, I have been struck by the similarity of the rhetoric to those who promote native species restoration today. They, too, were sure they were doing the right thing for the world.

And if you need more evidence for the old maxims about circles, about things changing and things staying the same, look

across the Pacific. China, today, is undergoing an industrial
revolution that in many ways mirrors the United States of the
nineteenth century. Pollution and environmental degradation
are rampant, wealth is being generated at a tremendous rate,
the growing middle class is becoming increasingly concerned
about losing a more elemental and presumably healthful way
of life. And one of the fastest growing sports around cities like
Beijing? Fishing for hatchery rainbows.[7]

Notes

Abbreviations

NACP: National Archives at College Park, College Park, Maryland
NARM: National Archives Rocky Mountain Region, Denver, Colorado
SIA: Smithsonian Institution Archives
UMMZ: Archives of the University of Michigan Museum of Zoology
USFC: United States Fish Commission

ONE. A Less Bold and Spirited Nation

1. Signor, *Southern Pacific*, 15–16.
2. Signor, *Southern Pacific*, 15–16.
3. Stone, "Report of Operations During 1872," 168–175.
4. Stone, "Report of Operations During 1872," 168–175; Hedgpeth, "Livingston Stone and Fish Culture," 131.
5. Stone, "Report of Operations During 1872," 173; Stone to Baird, 25 September 1879, NACP RG 22(117).
6. Hallock, *The Fishing Tourist*, 220; Farquhar, *Up and Down California*.
7. Farquhar, *Up and Down California*.
8. Stone, "Report of Operations During 1872," 173.
9. Woodham Smith, *The Reason Why*, 262; Marsh, "Artificial Propagation," 64–65.
10. Marsh, "Artificial Propagation," 64.
11. Marsh, "Artificial Propagation," 64–65.
12. Marsh, "Artificial Propagation," 68.
13. For example, *Forest and Stream*, "Vermont Game Laws," 297.
14. Marsh, "Artificial Propagation," 66.
15. Shanks, "Fish Culture"; Allard, *Spencer Fullerton Baird*, 111.
16. Guinet, "Remy et Gehin," 405 (translated by author).
17. Kinsey, "Seeding the Water"; Guinet, "Remy et Gehin"; Shanks, "Fish Culture," 721–739.
18. Bowen, "History of Fish Culture," 72.

TWO. Essentially a National Matter

1. Thompson, "The First Fifty Years," 1.
2. Stone, "Trout Culture," 52–58.

3. Shanks, "Fish Culture," 728; Stone, "Trout Culture," 48; Reiger, *American Sportsmen*, 22.

4. Allard, "Spencer Fullerton Baird," 116–121.

5. Henshall, "Black Bass," 505.

6. California Fish and Game, *Report of the Commissioners, 1879*; Sjovold, *An Angling People*, 116–117.

7. Smith, *Scaling Fisheries*, 43.

8. Allard, *Spencer Fullerton Baird*, 81–82.

9. Allard, *Spencer Fullerton Baird*, 94–101.

10. Allard, *Spencer Fullerton Baird*, 96–98.

11. Bowen, "History of Fish Culture," 77–83; USFC, *Annual Report, 1872–1873*, xvi.

12. Allard, *Spencer Fullerton Baird*, 122–126.

13. Roosevelt, *Fish Culture*, 4.

14. Roosevelt, *Fish Culture*, 3; *New York Times*, "Pisciculture," 28 April 1872.

15. Roosevelt, *Fish Culture*, 5.

16. Allard, *Spencer Fullerton Baird*, 130; Bowen, "History of Fish Culture," 82.

17. Bowen, "History of Fish Culture," 82–83; see also USFC, *Annual Report, 1872–1873*, xv, and 757–763.

18. USFC, *Annual Report, 1872–1873*, lxvi.

19. Rawson, *In Common with All Citizens*, 136.

20. Allard, *Spencer Fullerton Baird*, 137.

21. USFC, *Annual Report, 1872–1873*, lxvi, 762, xxiii; Bowen, "History of Fish Culture," 83.

22. Oakes, *Genealogical and Family History*, 626–633; Saunderson, *History of Charlestown*, 248.

23. Saunderson, *History of Charlestown*, 248–249.

24. Oakes, *Genealogical and Family History*, 628; *Utica Globe*, "Obituary: Livingston Stone," 4 January 1913; Caldwell, *The Last Crusade*, 48.

25. Saunderson, *History of Charlestown*, 249.

26. Stone, "Brief Reminiscences," 337–339.

27. Stone, "Trout Culture," 46–56; *The National Eagle* (Claremont, New Hampshire), "Charlestown," 6 June 1885; Shanks, "Fish Culture," 721–739; Hedgpeth, "Livingston Stone and Fish Culture," 128.

28. Stone, *Domesticated Trout*.

29. Bowen, "History of Fish Culture," 78; Allard, *Spencer Fullerton Baird*, 113–116; Stone to Willard Perrin, 14 January 1871, private collection.

30. *Cultivator and Country Gentleman* (Albany, New York), "Trout Breeding Property for Sale," 18 July 1872; Stone, "Report of Operations During 1872," 201.

31. Starr, *California*, 121, 80.
32. Stone, "Report of Operations During 1872," 168.
33. Stone, "Report of Operations During 1872," 169.
34. *Forest and Stream*, "United States Fish Hatching," 84.
35. Stone, "Report of Operations During 1872," 197–215.

THREE. Let the Best Fish Win

1. Wales, "General Report of Investigations on the McCloud," 272–309;
   Behnke, "Livingston Stone, J. B. Campbell," 20–22; Behnke, "First
   Documented Case of Anadromy," 582–585.
2. Stone, "Report of Operations During 1872," 200–201; *San Francisco
   Chronicle*, "Fish Culture," 12 December 1871, 4.
3. Osborne, "Acclimatizing the World," 135–151.
4. Osborne, "Acclimatizing the World," 137, 139.
5. Osborne, "Acclimatizing the World," 137, 146–147; *Forest and Stream*,
   "American Acclimatization Society," 305; *San Francisco Chronicle*,
   "Acclimatization," 11 February 1871.
6. *Forest and Stream*, "American Acclimatization Society," 305; Lever,
   Naturalized Birds, 4, 482.
7. USFC, *Annual Report, 1878*, xlv; McEvoy, *Fisherman's Problem*, 104;
   Dunlap, "Sport Hunting and Conservation," 318; *Forest and Stream*,
   "Rainbow Trout in England," 373.
8. Towle, "Authored Ecosystems," 58–60; Rinne and Stefferud, "Single
   Versus Multiple Species Management," 360.
9. Roosevelt, *Fish Culture*, 13.
10. Roosevelt, *Fish Culture*, 13.
11. *San Francisco Alta*, "Trout Culture," 22 May 1871.
12. *Herald of Gospel Liberty*, "Father of Fish Culture," 1 November 1883;
    Green, "Seth Green and Fish Culture."
13. Green, "Seth Green and Fish Culture."
14. *San Francisco Chronicle*, "Fish Culture," 12 December 1871.
15. *San Francisco Alta*, "Trout Culture," 22 May 1871; *San Francisco Alta*,
    "Pisciculture," 16 July 1873; Stone, "Report of Operations During
    1872," 200–201.
16. Stone, "Report of Operations During 1872," 200–201; Behnke, "First
    Documented Case of Anadromy," 582–585; *San Francisco Chronicle*,
    "Fish Culture," 12 December 1871.
17. *Forest and Stream*, "Fish Culture in New York."
18. New York Commissioners of Fisheries, *Report* (1879), 5, 44.
19. New York Commissioners of Fisheries, *Report* (1879), 5, 44; Green,
    "The Rainbow Trout."

FOUR. As Many Different States as Possible

1. Smiley, "A Statistical Review," 829; USFC, *Annual Report, 1888,* xxxv;
   Bowen, "History of Fish Culture," 81–85; Stone, "The Artificial
   Propagation of Salmon," 218–219.
2. Stone, "Report of Operations at the United States Trout Ponds, 1879,"
   716.
3. Stone, "Report of Operations at the United States Trout Ponds, 1879,"
   716.
4. Stone, "Report of Operations at the United States Trout Ponds, 1879,"
   717.
5. Allard, "Spencer Fullerton Baird," 290; Goode, "Status of the U.S.
   Fish Commission in 1884," 1157.
6. Leach, "Propagation and Distribution of Food Fishes, 1931," 684–685;
   Bowen, "History of Fish Culture," 88–89.
7. Baird to C. G. Atkins, 16 November 1877, NACP RG 22(15); Allard,
   Spencer Fullerton Baird, 290; Goode, "Status of the U.S. Fish Commission in 1884," 1157.
8. USFC, *Annual Report, 1882,* lxxiii.
9. USFC, *Annual Report, 1879,* xxx–xxxi; Stone to Baird, 27 October 1875,
   NACP RG 22(117); New York Commissioners of Fisheries, *Report*
   (1879), 5; California Fish and Game, *Report of the Commissioners of
   Fisheries of the State of California,* years 1877–1879.
10. Allard, *Spencer Fullerton Baird,* 127.
11. Stone, "Report of Operations at the Trout-Breeding Station, 1882,"
    853; Stone, "Report of Operations at the United States Trout Ponds,
    1879," 718.
12. Sjovold, *An Angling People,* 151.
13. Stone, "Report of Operations at the United States Trout Ponds, 1880,"
    598; Stone, "Report of Operations at the United States Salmon-
    Breeding Station, 1879," 699–700.
14. Stone, "Report of Operations at the United States Trout Ponds, 1879,"
    718.
15. Stone to Baird, 17 June 1884, NACP RG 22.
16. Stone to Baird, 15 April 1877, NACP RG 22(4); Stone, "Report of
    Operations at the United States Trout Ponds, 1879," 718.
17. Green, "Seth Green and Fish Culture."
18. USFC, *Annual Reports,* 1879–1887.
19. Taylor, *Making Salmon,* 84–85.
20. Stone, "Report of Operations at the McCloud River Trout Pond Sta-
    tion, 1888."
21. Stone to McDonald, 1 March 1890, NACP RG 22(117).

22. Seamans, "Mr Livingston Stone," 7–10; Taylor, *Making Salmon*, 84–85; Hedgpeth, "Livingston Stone and Fish Culture," 139.

23. Becky McCue (Stone's granddaughter), in discussion with author, 29 November 2007.

24. Yoshiyama and Fisher, "Long Time Past," 6–22.

25. Yoshiyama and Fisher, "Long Time Past," 6–22.

26. Jordan and Evermann, *American Food and Game Fishes*, 198. The New York Fish Commission did not commonly share its fish with other states. See, for example, New York Commissioners of Fisheries, *Report of the Commissioners* (1885), 9. R. J. Behnke, in "First Documented Case of Anadromy" and "Livingston Stone, J. B. Campbell," argues that it is erroneous to trace all present-day hatchery rainbows back to the McCloud USFC hatchery. I do not deny that today's hatchery rainbows descend from many different rainbow and steelhead stocks. However, given how widely Stone and the USFC dispersed the rainbows from the McCloud (to thirty-three states, according to the USFC Annual Reports from 1879–1887), and given how frequently fish culturists interbred their fish with those from other hatcheries (see Leitritz, *A History of California's Fish Hatcheries*), it also seems very likely that most hatchery rainbows today contain at least a few genes from Stone's McCloud operation.

FIVE. A New Variety of Trout

1. Behnke, *Native Trout of Western North America*; Behnke, *Trout and Salmon of North America*.

2. Behnke, *Native Trout of Western North America*, 13–21, 166–171.

3. Behnke, *Trout and Salmon of North America*, 69–70.

4. Behnke, *Trout and Salmon of North America*, 70–71.

5. Agassiz, *Studies on Glaciers*; Mayr, *Growth of Biological Thought*, 566; Wegener, *Origin of Continents*.

6. Steller, *Steller's History of Kamchatka*; Willmore and Engel, "Translators' Preface," ix–xiv.

7. Willmore and Engel, "Translators' Preface," xiii; Steller, *Steller's History of Kamchatka*, 114–115.

8. Behnke, *Trout and Salmon of North America*, 19; Rawson, *In Common with All Citizens*, 126.

9. Lewis and Clark, *Journals of Lewis and Clark*, 13 March 1806, 22 August 1805.

10. Smith and Stearley, "The Classification and Scientific Names," 5–6; Behnke, *Native Trout of Western North America*, 5.

11. Suckley, "North American Species of Salmon and Trout," 129; Smith

and Stearley, "The Classification and Scientific Names," 6; Stone to Baird, 27 October 1875, NACP RG 22(117); Stone to Baird, 18 November 1875, NACP RG 22(117); Behnke, *Native Trout of Western North America*, 162.

12. Smith and Stearley, "The Classification and Scientific Names," 4–10; Robins, Untitled sidebar, 5.
13. Gerald Smith, email to author, 30 September 2008.
14. Walton and Cotton, *Complete Angler*, 95; Grubic and Herd, "Astraeus Origins," 16–22.
15. Calabi, *Trout and Salmon*, 39; Raymond, *Steelhead Country*, 21–22; for more examples, see Rawson, *In Common with All Citizens*, 185.
16. Behnke, *Trout and Salmon of North America*, 14–17.
17. Behnke, *Trout and Salmon of North America*, 13.
18. American Fisheries Society, *Common and Scientific Names*, 210; Behnke, *Native Trout of Western North America*, 172; Behnke, *Trout and Salmon of North America*, 18, 19, 73, 105–114.
19. Behnke, *Native Trout of Western North America*, 161–172.

six. Define Me a Gentleman

1. Lubowski, *Major Uses of Land*.
2. Halverson, "Stocking Trends," 69–75; National Agricultural Statistics Service, *Trout Production*.
3. MacCrimmon, "World Distribution of Rainbow Trout," 663–704.
4. Colorado State Government, "Colorado State Archives Symbols and Emblems"; Utah State Government, "Utah State Fish"; Cambray, "Impact on Indigenous Species."
5. Tattersall, *Becoming Human*, 9; British Museum, *Spreadsheet*, 4.
6. Rawson, *In Common with All Citizens*, 63–64.
7. Rawson, *In Common with All Citizens*; Plutarch, *Antony*.
8. Dulles, *A History of Recreation*.
9. Court of Assistants of the Colony of the Massachusetts Bay, *Records*, 37; Dulles, *A History of Recreation*, 4; Cronon, *Changes in the Land*, 56.
10. Rawson, *In Common with All Citizens*, 92–93.
11. Continental Congress, *The Articles of Association: October 20, 1774*; Dulles, *A History of Recreation*, 63, 86; Nichols, *Forty Years*, 206.
12. Reiger, *American Sportsmen*, 45; Schullery, *American Fly Fishing*, 127–128.
13. *Forest and Stream*, "Trout Day"; *Forest and Stream*, "New Game Preserve"; Behnke, "America's First Brown Trout"; Smiley, "Brief Notes," 360; *New York Times*, "Cultivation of Trout," 15 March 1885.
14. Catholic World, "Sanitary and Moral Condition"; Mumford, *The City*, 468.

15. Sheehy, "American Angling," 79; Dulles, *A History of Recreation*, 211, 271; Reiger, *American Sportsmen*, 55–56.
16. Duncan, *Miles from Nowhere*, 4–6; Turner, "Significance of the Frontier."
17. Hallock, *The Fishing Tourist*, 224.
18. Schullery, *American Fly Fishing*, 131; La Motte, "Fish Propagation," 237; Leitritz, *A History of California's Fish Hatcheries*.
19. Bawden, *Reinventing the Frontier*, 85; Gross, *Changing Shape of the American Landscape*, 1.
20. Sjovold, *An Angling People*, 104; Reiger, *American Sportsmen*, 53.
21. Washabaugh and Washabaugh, *Deep Trout*, 80; Lears, *No Place of Grace*, 27.
22. Holmes, *Autocrat of the Breakfast-Table*, 881; Lears, *No Place of Grace*, 30, 47–58, 108.
23. Lears, *No Place of Grace*, 30, 47–58.
24. Bawden, *Reinventing the Frontier*, 245; Hallock, *The Fishing Tourist*, 68; Schullery, *American Fly Fishing*, 46; Roosevelt and Grinnell, *American Big-Game Hunting*, 14–15; *New York Times*, "Field Sports," 12 January 1874.
25. Nash, *Wilderness and the American Mind*, 60; Cronon, "Trouble with Wilderness," 73.
26. Lears, *No Place of Grace*, 147; Freud, *Civilization and Its Discontents*, 57.
27. Lears, *No Place of Grace*, xi–xii.
28. Lears, *No Place of Grace*.
29. Hallock, *The Fishing Tourist*, 25–26.
30. Sjovold, *An Angling People*, 117.
31. Sjovold, *An Angling People*, 114–115; Rawson, *In Common with All Citizens*, 115–116.
32. Behnke, "America's First Brown Trout"; Shafer, "Angling with the Bark On."
33. Stewart, "Twelve Years Experience"; Gordon, "Rainbow Trout."
34. *Forest and Stream*, "Wonderful Supply of Rainbow Trout."
35. *Forest and Stream*, "Forest Park Associations"; Reiger, *American Sportsmen*, 7; Sjovold, *An Angling People*, 5–12.
36. *Forest and Stream*, "Vermont Game Laws."
37. *Forest and Stream*, "Trout Poaching"; Knox, "Summer Clubs"; *Forest and Stream*, "Defiance of Law"; *Forest and Stream*, "Game Protection"; Sjovold, *An Angling People*, 200–202.
38. Reiger, *American Sportsmen*, 1–4; Taylor, *Making Salmon*, 202.

SEVEN. Paying Customers and Hatchery Product

1. Leach et al., "Propagation and Distribution of Food Fishes, 1939," 556; Roosevelt, *Fish Culture.*
2. Dollar and Katz, "Rainbow Trout Brood Stocks," 167–172.
3. Kincaid et al., *National Fish Strain Registry.*
4. Cheney, "Angling Notes"; Cobb, "Results of Trout Tagging"; Needham and Behnke, "Origin of Hatchery Rainbow Trout."
5. Wiley, "Common Sense Protocol," 468; Van Vooren, "The Roles of Hatcheries," 512–513.
6. Curtis Milliron, in email to author, 24 January 2008; C. Meyers, "Different Strains, Everyone Gains," *Denver Post,* 6 June 1997.
7. Bob Peters, in discussion with author, 22 September 2006.
8. Trefethen, *Crusade for Wildlife,* 158–159; Reiger, *American Sportsmen,* 93.
9. American Sportfishing Association, *Sportfishing in America;* Epifanio, "Status of Coldwater Fishery Management," 20; Trefethen, *Crusade for Wildlife,* 159–160; Rawson, *In Common with All Citizens,* 223–224.
10. Buchanan, *Public Finance.*
11. Goode, "Status of the U.S. Fish Commission in 1884," 1157.
12. Nielsen, "History of Inland Fisheries," 23.
13. *Sacramento Bee,* "UCD Develops Hybrid Trout That Get Bigger Quicker," 4 May 1969; C. Meyers, "Enough Not Enough for DOW; Four Divisions Struggle to Get Their Fair Share," *Denver Post,* 8 July 2001; C. Meyers, "Low Fish Stocks, Sales Put DOW in Dilemma," *Denver Post,* 16 September 2001.
14. Goode, "Status of the U.S. Fish Commission in 1884," 1157; Tunison et al., "Extended Survey of Fish Culture," 260–261; Tunison et al., "Survey of Fish Culture," 62–63; Hagen and O'Connor, "Public Fish Culture."
15. E. R. Vincent, in discussion with the author, 10 April 2008; U.S. Fish and Wildlife Service, *Economic Effects.*
16. Behnke, *Summary of Progress;* L. Shapovalov, "State Expert Explains Relationship Between Number of Trout Planted and Angler's Creel," *Sacramento Bee,* 19 July 1950; Williams, "Trout Are Wildlife"; Gabelhouse, Jr., "Staffing, Spending, and Funding."
17. Hazzard and Shetter, "Results from Experimental Plantings," 209; Leitritz, *A History of California's Fish Hatcheries,* 50.
18. Leitritz, *A History of California's Fish Hatcheries,* 52; Hazzard and Shetter, "Results from Experimental Plantings."
19. *New York Times,* "Pisciculture," 28 April 1872.
20. California Fish and Game, *Report for Years 1874 and 1875,* 18; Towle, *Authored Ecosystems.*

21. Muir, "Salmon-Breeding"; Leach, "Propagation and Distribution of Food Fishes, 1931," 684–685.
22. L. Shapovalov, "State Expert Explains Relationship Between Number of Trout Planted and Angler's Creel," *Sacramento Bee*, 19 July 1950, 29; Benson, "The American Fisheries Society," 16.
23. Whelan, "A Historical Perspective," 309.
24. Hoover and Johnson, "Migration and Depletion"; Tunison et al., "Survey of Fish Culture," 31–46.
25. Hazzard and Shetter, "Results from Experimental Plantings."
26. Nielsen, "History of Inland Fisheries," 21.
27. National Park Service Organic Act of 1916 (16 U.S.C. 1, 2, 3, and 4); Yellowstone National Park Act of 1872 (17 Stat. 32); Greer, "The United States Forest Service," 162; Whelan, "A Historical Perspective," 312.
28. Hazzard and Shetter, "Results from Experimental Plantings," 196, 205; Shoemaker et al., "Frederic Collin Walcott," 101.
29. Dulles, *A History of Recreation*, 204, 387; Bureau of Sport Fisheries and Wildlife, "Proposed Program of Extension."
30. Faist, *Air Operations*, 41.
31. Faist, *Air Operations*.
32. Faist, *Air Operations*.
33. Faist, *Air Operations*.
34. Gordon, "Now Trout Fly Too!"; R. Knapp, in discussion with author, 23 July 2006.
35. Faist, *Air Operations*.
36. J. Rendel, "Wood, Field and Stream," *New York Times*, 2 November 1945; Prevost, "Experimental Stocking"; Prevost and Piché, "Observations on the Respiration"; Saldana, "Aerial Fish Plant."
37. Faist, *Air Operations*.

EIGHT. A Full-Scale Military Operation

1. Fradkin, *A River No More*, 35.
2. Quartarone, *Historical Accounts*, 37.
3. Page and Burr, *Freshwater Fishes*; Quartarone, *Historical Accounts*; U.S. Fish and Wildlife Service, *Colorado Squawfish Recovery Plan*, 12.
4. Quartarone, *Historical Accounts*.
5. Binns et al., *Planning, Operation, and Analysis*; McEvoy, *Fisherman's Problem*, 104–105; Pister, "Wilderness Fish Stocking," 119–120; Reiger, *American Sportsmen*, 85.
6. Krukoff and Smith, "Rotenone Yielding Plants"; Lennon et al., "Reclamation of Ponds," 33.

7. Stickney, *Aquaculture in the United States,* 162, 226; Lennon et al., "Reclamation of Ponds."

8. Lennon et al., "Reclamation of Ponds"; Miller, "Underwater Life Worth Saving?"

9. Binns et al., *Planning, Operation, and Analysis,* 1.

10. H. S. Crane and D. Andriano to R. M. Stroud, 29 January 1962, NARM RG 22; Utah Department of Fish and Game and Bureau of Sport Fisheries and Wildlife, *Fish and Wildlife Values;* J. C. Gatlin to Director, Bureau of Sport Fisheries and Wildlife, 14 February 1962, NARM RG 22; Binns et al., *Planning, Operation, and Analysis,* 4; U.S. Department of the Interior, *Measures to Improve Sport Fishing.*

11. Binns et al., *Planning, Operation, and Analysis,* 1.

12. Binns et al., *Planning, Operation, and Analysis.*

13. G. Miller (son of R. R. Miller), in discussion with author, 19 March 2008; Cashner et al., "Robert Rush Miller."

14. G. Miller (son of R. R. Miller), in discussion with author, 19 March 2008; Cashner et al., "Robert Rush Miller."

15. Some of these letters can be found in the National Archives in College Park, Maryland. Many more are stored at the University of Michigan Museum of Zoology; Miller to R. E. Johnson, 12 December 1962, NACP RG 22(266).

16. Miller to A. H. Carhart, 29 October 1962, UMMZ.

17. G. A. Moore to Miller, 27 February 1963, UMMZ; H. K. Hagen to Miller, 5 February 1963, UMMZ; G. F. Edmunds to Miller, 21 November 1962, UMMZ; G. R. Smith to Miller, 18 May 1964, UMMZ; Miller to R. E. Johnson, 12 December 1962, UMMZ.

18. Miller to W. E. Stegner, 5 March 1962, UMMZ; Miller to C. Cottam, 17 October 1962, UMMZ.

19. S. J. Jiacoletti to American Society of Ichthyologists and Herpetologists, 28 July 1961, NARM RG 22; Holden, "Ghosts of the Green River"; C. L. Hubbs to R. Conant, 27 November 1962, UMMZ; J. C. Gatlin to Director, Bureau of Sport Fisheries and Wildlife, 14 February 1962, NARM RG 22.

20. J. E. Hemphill to Chief of Division of Sport Fisheries, 3 January 1962, NACP RG 22(266).

21. Quartarone, *Historical Accounts.*

22. Stone, "Tri-State Treatment."

23. Binns et al., *Planning, Operation, and Analysis.* 17; G. R. Smith, in discussion with author, 11 February 2008.

24. Binns et al., *Planning, Operation, and Analysis.* 14, 20; Sherer, "Mechanism of Toxicity in Rotenone"; Kamel et al., "Pesticide Exposure"; Brown et al., "Pesticides and Parkinson's Disease"; C. Queal, "They

Kill Fish to Make Fishing Better," *Denver Post,* 7 September 1962; J. R. Walton to Supervisor, Office of River Basin Studies, U.S. Fish and Wildlife Service, 25 July 1962, NARM RG 22; J. C. Gatlin to Director, Bureau of Sport Fisheries and Wildlife, 14 February 1962, NARM RG 22.

25. C. H. Bennett to S. J. Jiacoletti, 26 December 1963, NARM RG 22; Binns et al., *Planning, Operation, and Analysis,* 22.

26. Binns et al., *Planning, Operation, and Analysis,* 32–53; Holden, "Ghosts of the Green River," 50.

27. Binns et al., *Planning, Operation, and Analysis,* 32–53; Post, "Chemical Test for Rotenone."

28. Binns et al., *Planning, Operation, and Analysis,* 47–52.

29. Binns et al., *Planning, Operation, and Analysis,* 47–52; Miller to O. L. Wallis, 30 November 1962, UMMZ.

30. Fox, *The American Conservation Movement,* 281–287.

31. V. Adams, "'CBS Reports' Plans a Show on Rachel Carson's New Book," *New York Times,* 30 August 1962.

32. Holden, "Ghosts of the Green River," 50; L. R. Garlick to Regional Director, Region 2, Bureau of Sport Fisheries and Wildlife, 11 September 1963, NARM RG 22; D. R. Franklin to Superintendent, Dinosaur National Monument, 29 October 1963, NARM RG 22; Miller, "Underwater Life Worth Saving?"

33. Miller to H. K. Hagen, 30 October 1962, UMMZ; Miller, "Underwater Life Worth Saving?"

34. R. L. Carson to R. E. Johnson, 6 February 1963, NACP RG 22(266); A. V. Tunison to C. Wolf, 22 April 1966, NACP RG 22(268); Holden, "Ghosts of the Green River," 51.

35. W. Sullivan, "Altering Nature: Vast Projects Affecting Wildlife in West Stir Controversy," *New York Times,* 22 September 1963; W. Sullivan, "Native Fish Are Killed to Help New Trout," *New York Times,* 11 August 1963.

36. Udall, *The Quiet Crisis;* Endangered Species Preservation Act of 1966 (Pub. L. No. 89–669, §§ 1–3, 80 Stat. 926 [repealed 1973]); Bean and Rowland, *Evolution of National Wildlife Law,* 194.

37. W. Stegner to Miller, 9 December 1962, UMMZ.

38. Udall, "Review of Green River Eradication"; A. V. Tunison, "Memorandum, 7 June 1963," NARM RG 22.

39. U.S. Department of Interior, "Interior Department Steps Up Fight"; Miller, "Underwater Life Worth Saving?" 7; U.S. Department of Interior, "Janzen to Develop Program"; Endangered Species Act of 1973 (Pub. L. No. 93–205, 87 Stat. 884 [1973]), current version: 16 U.S.C. §§ 1531–1543; Bean and Rowland, *Evolution of National Wildlife Law;*

U.S. Fish and Wildlife Service, "Committee Information Sheet"; Miller to S. Udall, 3 June 1963, UMMZ; W. King to Regional Director, Bureau of Sport Fisheries and Wildlife, Atlanta, 8 April 1965, NACP RG 22(268); Bureau of Sport Fisheries and Wildlife, *Rare and Endangered Fish and Wildlife.*

40. C. Queal, "They Kill Fish to Make Fishing Better," *Denver Post,* 7 September 1962.

41. *Denver Post,* "Big Fish Poison Job Slated for Wyoming," 2 September 1962.

42. *Salt Lake Tribune,* "Fish Treatment Opens Tuesday," 2 September 1962; Quartarone, *Historical Accounts;* Regenthal, "Treatment Complete"; Wagner, "Mission Accomplished."

43. Wyoming Game and Fish Commission, "Suggested Editorial," 22 June 1962.

44. U.S. Fish and Wildlife Service, "Endangered Species Program"; T. Hartman, "Fish Story," *Rocky Mountain News,* 4 December 2000.

45. Wiley, "Man and Native Fishes."

46. Wiley, "Man and Native Fishes"; F. P. Briggs to M. Ellis, 1 March 1962; J. C. Gatlin to Director, BSFW, 22 August 1961, NARM RG 22.

47. G. R. Smith, email to author, 11 February 2008.

48. U.S. Fish and Wildlife Service, *Briefing Statement on Flaming Gorge;* J. C. Gatlin to Director, Bureau of Sport Fisheries and Wildlife, 14 February 1962, NARM RG 22; A. V. Tunison to C. Wolf, 22 April 1966, NACP RG 22(268). H. S. Crane and P. A. Dotson (Director of the Utah Department of Fish and Game, and Acting Regional Coordinator) to G. R. Smith, 10 November 1960, UMMZ, declares: "At the most, I do not expect the reservoir to be more than marginal habitat for carp . . . the important thing in the initial stage of our stocking . . . is to eliminate all competition from existing fish species." For exceptions, see W. F. Sigler to R. M. Bailey, 7 May 1961, UMMZ, which states that the dams will probably deplete the native fish no matter what and that the Utah Fish and Game Department is very sympathetic to those who want to "perpetuate native species regardless of whether they are considered as game."

49. W. F. Sigler to R. M. Bailey, 7 May 1961, UMMZ; Miller to O. L. Wallis, 29 May 1961, UMMZ; U.S. Fish and Wildlife Service, "Endangered Colorado River Basin Fish"; H. Tyus, in discussion with author, 1 November 2005; G. R. Smith, in discussion with author, 11 February 2008.

50. Utah State Division of Wildlife Resources and Wyoming Game and Fish Commission, *Green River and Flaming Gorge Reservoir;* Holden, "Ghosts of the Green River," 52–54.

NINE. Money Makes a Way

1. Schullery, *American Fly Fishing*, 190–208, 247; Griffith, *For the Love of Trout*, 130–132; Schley, "Somewhere a River Begins," 213–214; Annin, "The Rainbow Trout"; Hummel, *Hunting and Fishing*, 52–53.

2. Griffith, *For the Love of Trout*, 89, 104; Behnke, *Trout and Salmon of North America*, 277.

3. Griffith, *For the Love of Trout*, 131; *Trout*, "A Trout Unlimited Retrospective."

4. Griffith, *For the Love of Trout*, 178.

5. Washabaugh and Washabaugh, *Deep Trout*, 122; Griffith, *For the Love of Trout*, 186–187.

6. *Trout*, "A Trout Unlimited Retrospective"; Behnke, "The First Forty Years."

7. Griffith, *For the Love of Trout*, 145–159.

8. Behnke, "Limit Your Kill."

9. Griffith, *For the Love of Trout*, 144, 154.

10. Griffith, *For the Love of Trout*, 137, 147, 131.

11. Whelan, "A Historical Perspective"; Griffith, *For the Love of Trout*, 131.

12. Whitney, "The Changing Role"; Behnke, "Limit Your Kill."

13. E. R. Vincent, in discussion with the author, 10 April 2008; *Montana Outdoors*, "Why Montana Went Wild."

14. Vincent, "The Madison River."

15. E. R. Vincent, in discussion with the author, 10 April 2008.

16. *Montana Outdoors*, "Why Montana Went Wild"; Whitney, "The Changing Role."

17. E. R. Vincent, in discussion with the author, 10 April 2008.

18. Whitney, "Who Pays for What," 13; Wells, "Wild Trout Management."

19. Wells, "Wild Trout Management."

20. Wells, "Wild Trout Management"; E. R. Vincent, in discussion with the author, 10 April 2008.

21. E. R. Vincent, in discussion with the author, 10 April 2008.

22. *Montana Outdoors*, "Why Montana Went Wild."

23. Annin, "The Rainbow Trout."

24. Leopold, *Game Management*, 394; Cobb, "Results of Trout Tagging"; see Vincent, "The Catchable Trout," for an example of an angler who believed stocking was responsible for deteriorating fishing in 1960; see *Montana Outdoors*, "Why Montana Went Wild," for examples of other anglers' complaints.

25. Mather, "Progress in Fish Culture"; Forbes, "The Investigation of a River," 182; Cobb, "Pacific Salmon Fisheries," 493; E. Neal, "In Search

of Elusive Trout," *San Francisco Examiner and Chronicle,* 6 February 1977.

26. Tunison, "Survey of Fish Culture," 31; Smith and Needham, "Problems Arising"; Hazzard and Shetter, "Results from Experimental Plantings"; Bogue, *Fishing the Great Lakes,* 305; Heidinger, "Stocking for Sport Fisheries Enhancement," 309–310; Taylor, *Making Salmon,* 203; "Rainbows in the Ozarks" (1963), NACP NAIL 22.82; "Glen Canyon Dam" (1965), NACP NAIL 115.71; "Cooperative Fish Culture" (1927), NACP NAIL 22.22.

27. Behnke, "Trading Stubbornness for Science."

28. American National Election Studies, "ANES Guide to Public Opinion."

29. Barbour, "Ecological Fragmentation."

30. Trout Unlimited, "Annual Report, 2006."

31. *Trout,* "A Trout Unlimited Retrospective"; Epifanio and Nickum, *Fishing for Answers.*

32. *Spokane Spokesman-Review,* "Field Reports," 1 August 2004; American Wildlands et al., *Petition;* U.S. Fish and Wildlife Service, "12-Month Finding for an Amended Petition"; *American Wildlands v. Norton,* 193 F. Supp. 2d 244 (D.D.C., 2002); U.S. Fish and Wildlife Service, "Reconsidered Finding."

33. Rawson, *In Common with All Citizens,* 238–304; Washabaugh and Washabaugh, *Deep Trout;* Munday, "Rivers of Our Discontent."

34. Montana Natural Streambed and Land Protection Act of 1975 (Title 75, Chapter 7, MCA); Montana Water Use Act of 1973 (Title 85, Chapter 2, MCA); Wells, "Wild Trout Management."

35. American Sportfishing Association, *Sportfishing in America.*

36. White et al., "Better Roles for Fish Stocking"; California Fish and Game, "DFG Adopts Strategic Plan"; Stone, "Fish Stocking Programs in Wyoming"; Van Vooren, "The Roles of Hatcheries"; C. Meyers, "State's Trout Program Can't Be a Meat Market," *Denver Post,* 15 March 1998, C2; Colorado Wildlife Commission, "Wild and Gold Medal Trout Management."

37. Leopold et al., "Wildlife Management"; Williams, "Hatchery Narcosis."

38. National Fish Hatchery Review Panel, *Report;* Williams, "Hatchery Narcosis"; U.S. General Accounting Office, "Classification of the Distribution of Fish"; U.S. General Accounting Office, "Authority Needed"; Halverson, "Stocking Trends."

TEN. The Way of the Passenger Pigeon

1. Walker and Nehring, "An Investigation to Determine the Cause(s)," A-108; Bartholomew and Reno, "History and Dissemination of Whirling Disease."
2. Bartholomew and Wilson, *Whirling Disease: Reviews.*
3. Wagner, "Whirling Disease Prevention"; Potera, "Fishing for Answers."
4. Harris, "Evolution of Colorado's Stocking Procedures."
5. Bartholomew and Reno, "History and Dissemination of Whirling Disease."
6. Bartholomew and Reno, "History and Dissemination of Whirling Disease."
7. E. Dentry, "Trout Takes Whirling Disease in Stride," *Rocky Mountain News,* 19 February 2002.
8. Meyer, "Aquaculture Disease."
9. Walker and Nehring, "An Investigation to Determine the Cause(s)," xvii, 82, A-124.
10. Nehring, "Biological, Environmental, and Epidemiological Evidence."
11. Walker and Nehring, "An Investigation to Determine the Cause(s)"; R. B. Nehring, in discussion with author, 27 September 2006.
12. R. B. Nehring, in discussion with author, 27 September 2006; Nehring, "Biological, Environmental, and Epidemiological Evidence."
13. Nehring, "Biological, Environmental, and Epidemiological Evidence."
14. Nehring, "Biological, Environmental, and Epidemiological Evidence"; Harris, "Evolution of Colorado's Stocking Procedures."
15. J. Kohler, "New State Policy Aimed at Eradicating Whirling Disease," *Associated Press,* 6 July 2001; C. Meyers, "Stop the Stocking of Infected Fish," *Denver Post,* 18 April 2000; B. Saile, "DOW Calls Moratorium on Stream-Stocking Plans," *Denver Post,* 16 April 1995; Walker and Nehring, "An Investigation to Determine the Cause(s)," 84.
16. B. Saile, "What's at End of Rainbow? Whirling Disease Requires Serious Studies," *Denver Post,* 1 May 1994; B. Saile, "Whirling Disease Fears Weren't So Reactionary," *Denver Post,* 11 December 1994.
17. D. Whipple, "Whirling Disease Widespread—Biologists Unsure Why Some Areas Suffer Huge Fish Losses," *Idaho Falls Post Register,* 11 February 1996; B. Saile, "Hole No Cure for Diseased Trout," *Denver Post,* 25 April 1995; C. Meyers, "Diseased Trout Remain on Hook for Late Reprieve," *Denver Post,* 30 April 1995; M. Obmascik, "Trout Quaran-

tine Pursued in Legislature," *Denver Post*, 20 April 1995; C. Meyers, "DOW, Trout Unlimited Feuding," *Denver Post*, 24 October 1999.

18. C. Meyers, "State's Trout Program Can't Be a Meat Market," *Denver Post*, 15 March 1998.

19. Nehring, "Stream Fisheries Investigations . . . F-237-R6"; E. Dentry, "Trout Takes Whirling Disease in Stride," *Rocky Mountain News*, 19 February 2002; C. Meyers, "Stocking Policy Hurting Spread of 'Clean' Trout," *Denver Post*, 17 February 2004; C. Meyers, "Fish-Stocking Policy Will Hamper Anglers," *Denver Post*, 13 November 2001; J. Kohler, "New State Policy Aimed at Eradicating Whirling Disease," *Associated Press*, 6 July 2001; Nehring, "Stream Fisheries Investigations . . . F-237-R11."

20. B. Saile, "Salmon, Trout at Risk of Disease," *Denver Post*, 8 April 1994; R. B. Nehring, in discussion with author, 27 September 2006; Nehring, "Biological, Environmental, and Epidemiological Evidence."

21. Vincent, "Relative Susceptibility of Various Salmonids."

22. P. Hofer to M. El-Matbouli, 10 November 2005, copied in email to author, 12 August 2008; El-Matbouli et al., "Identification of a Whirling Disease Resistant Strain"; El-Matbouli in email to author, 12 August 2008; Colorado Division of Wildlife, "New Strains of Rainbow Trout"; C. Meyers, "Rainbow Troubles May Have Relief," *Denver Post*, 17 February 2002.

23. Doyle et al., "Use of DNA Microsatellite Polymorphism"; White et al., "Better Roles for Fish Stocking"; Behnke, "Nature and Nurture"; Behnke, "Tale of Two Rivers."

24. Allendorf et al., "Whirling Disease and Wild Trout"; Miller and Vincent, "Rapid Natural Selection for Whirling Disease."

25. E. R. Vincent, in discussion with the author, 10 April 2008.

26. E. R. Vincent, in discussion with the author, 10 April 2008.

27. *San Jose Mercury News*, "Wolf-Reintroduction Project Dealt Blow," 29 July 1995.

28. Colorado Division of Wildlife, "New Strains of Rainbow Trout."

29. C. Meyers, "Uncertainties Clouding Hatchery System Future," *Denver Post*, 18 March 2001; C. Meyers, "Hot Debate Merely Causes Delay in Two-Fish Mandate," *Denver Post*, 19 September 1997.

30. C. Meyers, "Low Fish Stocks, Sales Put DOW in Dilemma," *Denver Post*, 16 September 2001; C. Meyers, "Number of Trout-Loving Anglers Doesn't Add Up," *Denver Post*, 14 August 2001; C. Meyers, "Fishermen Evolve as Stocking Shifts; Fingerlings Set Changes," *Denver Post*, 1 June 2003.

31. C. Meyers, "State's Trout Program Can't Be a Meat Market," *Denver*

Post, 15 March 1998; M. Obmascik, "Spawning Disaster," *Denver Post,* 16 April 1995.

32. C. Meyers, "DOW Loses Activist as Kochman Set to Retire," *Denver Post,* 7 April 2002.

33. I had private conversations with CDOW personnel who will remain nameless. But for examples where the tension has spilled into the press, see K. Licis, "Whirling Disease Is Out of the Bottle in Colorado," *Colorado Springs Gazette,* 2 June 2000; C. Meyers, "Stocking Policy Hurting Spread of 'Clean' Trout," *Denver Post,* 17 February 2004.

ELEVEN. A Single New Mongrel Species

1. Stoddard et al., "Ecological Assessment of Western Streams."
2. Amy Ackerman, in discussion with author, 29 October 2008.
3. Schade and Bonar, "Distribution and Abundance of Nonnative Fishes"; Tyus and Saunders, "Nonnative Fish Control."
4. Williams et al., "Fishes of North America Endangered . . . 1989"; Allan and Flecker, "Biodiversity Conservation in Running Waters"; Richter et al., "Threats to Imperiled Freshwater Fauna"; Jelks et al., "Conservation Status"; Miller et al., "Extinctions of North American Fishes"; Saunders et al., "Freshwater Protected Areas."
5. Miller et al., "Extinctions of North American Fishes."
6. Blinn et al., "Effects of Rainbow Trout Predation"; Kruse et al., "Status of Yellowstone Cutthroat"; Hearn, "Interspecific Competition and Habitat Segregation."
7. See, for example, Whelan, "A Historical Perspective."
8. Hazzard and Shetter, "Results from Experimental Plantings"; Li and Moyle, "Management of Introduced Fishes."
9. See, for example, Nielsen, "History of Inland Fisheries," 21.
10. Lewis and Clark, *Journals of the Lewis and Clark Expedition,* 13 June 1805; Behnke, *Trout and Salmon of North America,* 158–159.
11. Behnke, *Trout and Salmon of North America,* 139–148; Behnke, *Native Trout of Western North America,* 14–21.
12. Behnke, *Trout and Salmon of North America,* 155–162; Lewis and Clark, *Journals of the Lewis and Clark Expedition,* 13 June 1805.
13. Liknes and Graham, "Westslope Cutthroat in Montana"; U.S. Fish and Wildlife Service, *Status Review for Westslope Cutthroat.*
14. Hauer et al., "Pattern and Process."
15. McCloud River rainbows were shipped to a U.S. Fish Commission hatchery in Wytheville, Va., in 1882 (USFC, *Annual Report, 1882,*

lxxii). In 1899, the Bozeman hatchery received rainbow trout from "different points." Most likely they came from Wytheville, Va., or Neosho, Mo. But since the Neosho rainbows originally came from Wytheville (USFC, *Annual Report, 1889–1891*, 46), all of the Bozeman fish can be traced to McCloud River rainbows (de C. Ravenal, "Report on Propagation and Distribution of Food-Fishes, 1899," xlii, lxxxix, lxvii). In 1903, twenty-five hundred rainbows were stocked in Laws Lake near Kalispell, Mont., almost certainly originating at the Bozeman hatchery (Titcomb, "Report on Propagation and Distribution, 1903," 43). In 1904, more rainbows were stocked in Knights Lake, also near Kalispell (Bowers, *Report of the Bureau of Fisheries 1904*, 46); Hitt et al., "Spread of Hybridization"; Liknes and Graham, "Westslope Cutthroat in Montana."

16. Hitt et al., "Spread of Hybridization."

17. Allendorf et al., "Intercrosses and the U.S. Endangered Species Act."

18. For example, Liknes and Graham, "Westslope Cutthroat in Montana"; Madsen, "Protection of Native Fishes."

19. Rubidge and Taylor, "Hybrid Zone Structure"; Hitt et al., "Spread of Hybridization"; Allendorf et al., "Intercrosses and the U.S. Endangered Species Act."

20. Tol and French, "Status of a Hybridized Population"; Martinez, "Identification and Status of Colorado River Cutthroat"; Cutter, "Gorgeous Goldens"; Don Thompson, "Feds, State Plan $1.3 Million Program," Associated Press, September 17, 2004; Cordes et al., "Identifying Introgressive Hybridization."

21. Allendorf and Leary, "Conservation and Distribution of Genetic Variation"; Madsen, "Protection of Native Fishes"; U.S. Fish and Wildlife Service, "Reconsidered Finding."

22. Allendorf and Leary, "Conservation and Distribution of Genetic Variation."

23. Allendorf and Leary, "Conservation and Distribution of Genetic Variation"; Campton and Kaeding, "Westslope Cutthroat Trout, Hybridization"; Rubidge and Taylor, "Hybrid Zone Structure"; Leary et al., "Hybridization and Introgression"; Muhlfeld et al., "Hybridization Rapidly Reduces Fitness."

24. Leary et al., "Hybridization and Introgression"; Currens, "Introgression and Susceptibility to Disease."

25. Allendorf and Leary, "Conservation and Distribution of Genetic Variation," 181.

26. Petersen, *The Modern Ark*, 72–73; Rhymer and Simberloff, "Extinction by Hybridization and Introgression."

27. O'Brien and Mayr, "Bureaucratic Mischief"; U.S. Fish and Wildlife Service and National Oceanic and Atmospheric Administration, "On the Treatment of Intercrosses"; Allendorf et al., "Intercrosses and the U.S. Endangered Species Act."

28. U.S. Fish and Wildlife Service and National Oceanic and Atmospheric Administration, "On the Treatment of Intercrosses"; Campton and Kaeding, "Westslope Cutthroat Trout, Hybridization."

29. American Wildlands et al., *Petition;* U.S. Fish and Wildlife Service, "12-Month Finding for an Amended Petition"; Caldwell, *The Endangered Species Act.*

30. *American Wildlands v. Norton,* 193 F. Supp. 2d 244 (D.D.C., 2002); U.S. Fish and Wildlife Service, "Reconsidered Finding"; *American Wildlands v. Norton* no. 05–1043 (D.D.C., 2007).

31. Campton and Kaeding, "Westslope Cutthroat Trout, Hybridization"; Allendorf et al., "Cutthroat Trout Hybridization and the U.S. Endangered Species Act."

32. F. Allendorf, in discussion with author, 9 April 2008.

33. Behnke, "Limit Your Kill."

34. Soon et al., "Unconscious Determinants of Free Decisions."

35. R. C. Archibold, "Helen Chenoweth-Hage, 68, Former Representative, Dies," *New York Times,* 4 October 2006; J. E. Yang, "Quotable," *Washington Post,* 5 November 1996, A9.

TWELVE. It Doesn't Do Any Good

1. Muir, *The Mountains of California,* 4–5.

2. Muir, *The Mountains of California,* chap. 6; Pister, "Wilderness Fish Stocking."

3. C. Milliron, in discussion with author, 15 May 2007; *San Francisco Alta,* "Ornithological and Piscatorial Acclimatizing Society," 20 April 1871; Brownlow, "Golden Trout Planting During 1928."

4. Brownlow, "Fish Planting in the High Sierra in 1929."

5. Leitritz, *A History of California's Fish Hatcheries,* 1970; Bahls, "Status of Fish Populations."

6. Muir, *Our National Parks,* chap. 6.

7. Grinnell and Storer, *Animal Life in the Yosemite,* 664; Pister, "Wilderness Fish Stocking."

8. Bahls, "Status of Fish Populations."

9. Pister, "The Desert Fishes Council."

10. Landres et al., "The Wilderness Act and Fish Stocking"; Wilderness Act of 1964 (16 U.S.C. 1131–1136, 78 Stat. 890).

11. Grinnell and Storer, *Animal Life in the Yosemite;* D. Perlman, "Cal Biologist Solves Mystery of Disappearing Sierra Frogs," *San Francisco Chronicle,* 13 May 2004.

12. Knapp, "Non-Native Trout in Natural Lakes," 13–14; Bradford et al., "Isolation of Remaining Populations"; C. T. Hall, "Fighting for the Frogs; Researchers Hope to Keep Amphibians from Disappearing in High Sierra," *San Francisco Chronicle,* 4 December 2000.

13. Sierra Nevada Aquatic Research Lab, "Sierra Nevada Aquatic Research Lab."

14. Blaustein et al., "Pathogenic Fungus Contributes to Amphibian Losses"; Stebbins, *Western Reptiles and Amphibians,* 86–87; Knapp, "Non-Native Trout in Natural Lakes," 14.

15. Bradford et al., "Isolation of Remaining Populations"; Knapp, "Non-Native Trout in Natural Lakes," 14.

16. Knapp, "Presentation at American Museum of Natural History Freshwater Biodiversity Conference."

17. Tunison et al., "Extended Survey of Fish Culture."

18. S. Elliott, "Sierra Nevada Species Dying: Agencies Working to Save Ecosystem of 'the Crown Jewel,'" *Modesto Bee,* 22 October 2000; Milliron et al., *Aquatic Biodiversity Management Plan for the West Walker Basin;* Curtis Milliron, in discussion with author, 15 May 2007.

19. Curtis Milliron, in discussion with author, 15 May 2007; R. A. Knapp, in discussion with author, 13 March 2006.

20. H. Abramson, "Pilots Score When Fingerlings Hit Lakes in the High Sierra," *Sacramento Union,* 19 September 1976.

21. Miller and Pister, "Management of the Owens Pupfish"; E. P. Pister, in discussion with author, 27 July 2006.

22. Pister, "The Desert Fishes Council," 56.

23. Pister, "The Desert Fishes Council," 56.

24. E. P. Pister, in discussion with author, 27 July 2006.

25. Pister, "The Desert Fishes Council," 63.

26. California Department of Fish and Game, "Budget Fact Book."

27. Rawson, *In Common with All Citizens;* R. Thornberry, "F and G Officials Poison Trout—In Effort to Repopulate Waters with Cutthroat, Biologists Eliminate Non-Native Rainbows and Brook Trout," *Idaho Falls Post Register,* 6 October 1999; Varley and Schullery, "Yellowstone Fishes"; Moore et al., "Restoration of Sams Creek"; Rosenlund et al., "Fisheries and Aquatic Management, Rocky Mountain National Park."

28. Lee, "Contribution of Nonnative Fish"; E. P. Pister, in discussion with author, 27 July 2006.

29. Milliron et al., *Aquatic Biodiversity Management Plan for the West Walker Basin.*
30. Rahel, "Homogenization of Fish Faunas."

EPILOGUE. The Last Generation of Troutfishers

1. Denton, *Rocky Mountain Radical;* Jordan and Evermann, *American Food and Game Fishes,* 209.
2. B. Frazee, "Bulked-up Rainbows in Future?" *Rutland Herald,* 4 June 2006; J. Drape, "McGwire Admits Taking Controversial Substance," *New York Times,* 22 August 1998.
3. Konrad and Konrad, "Rainbow Trout World Record"; Chandler, "Pending World Record."
4. S. Chan, "A School of Fish: Idaho Teaches Trout How to Eat Worms," *Wall Street Journal,* 15 August 1996.
5. Gay, "Rainbow Trout Planting."
6. Hyatt et al., *Trout Management Plan;* U.S. General Accounting Office, "Classification of the Distribution of Fish."
7. H. W. French, "A Novel, by Someone, Takes China by Storm," *New York Times,* 3 November 2005; Zhaoming and Yuhui, "Cold Water Fish Culture in China."

Bibliography

Many of the sources for this book are now available online, especially if you have access to a university library. The reports of the U.S. Fish Commission, including Livingston Stone's wonderful descriptions of his expeditions to California, are currently available to the public through the National Oceanic and Atmospheric Administration's online library. Likewise, many old copies of *Forest and Stream*, which provides a fascinating window on fishing and society in the nineteenth century, are now available through ProQuest's American Periodical Series Online. Many old books and periodicals can also be searched and read in full through Google Book Search. Lexis-Nexis and other newspaper archives also provide a wonderful resource.

Many of the letters and other sources reside in archives and private collections. The Shasta Historical Society, the Charlestown Historical Society, and the Rochester Museum and Science Center all hold valuable records pertaining to Livingston Stone, Seth Green, and the early days of fish culture in the United States. Many of Stone's private letters and photographs are held by Becky McCue, his granddaughter—my thanks again to her for allowing me to look at them. The Dowling College Library holds photos, documents, and memorabilia from the South Side Sportsmen's Club.

The National Archives in College Park, Maryland (NACP), holds thousands of letters and other documents from the U.S. Fish Commission and the U.S. Fish and Wildlife Service. Spencer Fullerton Baird kept catalogued copies of almost every letter he wrote and received, including many by Stone, and many of them reside there. Many internal memoranda and letters relating to the Green River rotenone project are at the National Archives in College Park and in the Rocky Mountain Region offices in Denver, Colorado (NARM). To help in finding them, I have tried to include the record group (RG) and entry number in references to these archives. The Smithsonian Institution Archives (SIA) in Washington, D.C., has many documents pertaining to Baird and to the Committee on Rare and Endangered Wildlife Species that was active in the 1960s. Many of the letters from and to Robert R. Miller and other documents relating to the Green River rotenone project are in the archives of the University of Michigan Museum of Zoology (UMMZ), where Jerry Smith went well beyond the call of duty to make them available to me.

Adams, S. B., C. A. Frissell, and B. E. Rieman. "Geography of Invasion in
    Mountain Streams: Consequences of Headwater Lake Fish Introduc-
    tions." *Ecosystems* (2001): 296–307.
Agassiz, L. *Studies on Glaciers: Preceded by the Discourse of Neuchâtel.*
    Translated and edited by A. V. Carozzi. New York: Hafner, 1967.
Allan, J. D., and A. S. Flecker. "Biodiversity Conservation in Running
    Waters." *BioScience* 43 (1993): 32–42.
Allard, D. C., Jr. *Spencer Fullerton Baird and the U.S. Fish Commission.* New
    York: Arno, 1978.
Allendorf, F. W. "Ecological and Genetic Effects of Fish Introductions—
    Synthesis and Recommendations." *Canadian Journal of Fisheries and
    Aquatic Sciences* 48, suppl. 1 (1991): 178–181.
Allendorf, F. W., and R. F. Leary. "Conservation and Distribution of Ge-
    netic Variation in a Polytypic Species, the Cutthroat Trout." *Conserva-
    tion Biology* 2, no. 2 (1988): 170–184.
Allendorf, F. W., R. E. Leary, N. P. Hitt, K. L. Knudsen, M. C. Boyer, and
    P. Spruell. "Cutthroat Trout Hybridization and the U.S. Endangered
    Species Act: One Species, Two Policies." *Conservation Biology* 19, no. 4
    (2005): 1326–1328.
Allendorf, F. W., R. F. Leary, N. P. Hitt, K. L. Knudsen, L. L. Lundquist,
    and P. Spruell. "Intercrosses and the U.S. Endangered Species Act:
    Should Hybridized Populations Be Included as Westslope Cutthroat
    Trout?" *Conservation Biology* 18, no. 5 (2004): 1203–1213.
Allendorf, F. W., P. Spruell, and F. M. Utter. "Whirling Disease and Wild
    Trout: Darwinian Fisheries Management." *Fisheries* 26, no. 5 (2001):
    27–29.
American Fisheries Society. *Common and Scientific Names of Fishes from the
    United States, Canada, and Mexico.* 6th ed., American Fisheries Society
    Special Publication 29. Bethesda: American Fisheries Society, 2004.
American National Election Studies. "The ANES Guide to Public Opin-
    ion and Electoral Behavior." American National Election Studies,
    www.electionstudies.org/nesguide/toptable/tab5a_1.htm (accessed
    3 December 2008).
American Society of Ichthyologists and Herpetologists. "Resolution of
    the American Society of Ichthyologists and Herpetologists." 3 April
    1961. NARM RG22.
American Sportfishing Association. *Sportfishing in America.* American
    Sportfishing Association, 2002, www.fishamerica.org/images/support/
    fish_eco_impact.pdf.
American Wildlands, Clearwater Biodiversity Project, Idaho Watersheds
    Project, Inc., Montana Environmental Information Center, Pacific
    Rivers Council, Trout Unlimited, Madison-Gallatin Chapter, B. Lilly.

Petition for a Rule to List the Westslope Cutthroat Trout (Oncorhynchus clarki lewisi) as Threatened Throughout Its Range. Washington: U.S. Fish and Wildlife Service, 1997.

Annin, J., Jr. "The Rainbow Trout." *Transactions of the American Fisheries Society* 11, no. 1 (1882): 20–24.

ASA Bulletin. "Hatchery Review Draws Fire from Fisheries Community." *ASA Bulletin* 465 (1995): 1–3.

Atkins, C. "Fish Culture." *American Naturalist* 1, no. 6 (1867): 296–304.

Bahls, P. "The Status of Fish Populations and Management of High Mountain Lakes in the Western United States." *Northwest Science* 66, no. 3 (1992): 183–193.

Bakke, B. "GAO Study Finds Feds Overstate Hatchery Benefits." *NW Fishletter,* 3 November 1999.

Barbour, M. G. "Ecological Fragmentation in the Fifties." In *Uncommon Ground; Rethinking the Human Place in Nature,* edited by W. Cronon, 233–255. New York: W. W. Norton, 1996.

Bartholomew, J. L., M. Mattes, M. El-Matbouli, T. S. McDowell, and R. P. Hedrick. "Susceptibility of Rainbow Trout Resistant to *Myxobolus cerebralis* to Selected Salmonid Pathogens." In *Propagated Fish in Resource Management,* American Fisheries Society Symposium 44, edited by M. J. Nickum, P. M. Mazik, J. G. Nickum, and D. D. MacKinlay, 549–557. Bethesda: American Fisheries Society, 2004.

Bartholomew, J. L., and P. W. Reno. "The History and Dissemination of Whirling Disease." In *Whirling Disease: Reviews and Current Topics,* American Fisheries Society Symposium 29, edited by J. L. Bartholomew and J. C. Wilson, 3–24. Bethesda: American Fisheries Society, 2002.

Bartholomew, J. L., and J. C. Wilson, eds. *Whirling Disease: Reviews and Current Topics,* American Fisheries Society Symposium 29. Bethesda: American Fisheries Society, 2002.

Bawden, T. *Reinventing the Frontier: Tourism, Nature, and Environmental Change in Northern Wisconsin, 1880–1930.* Ph.D. diss., University of Wisconsin, Madison, 2001.

Bean, M. J., and M. J. Rowland. *The Evolution of National Wildlife Law.* 3rd ed. Westport, Conn.: Praeger.

Behnke, R. J. "About Trout." *Trout,* Spring 2003, 54–56.

Behnke, R. J. "About Trout: America's First Brown Trout." *Trout,* Summer 2003, 55–57.

Behnke, R. J. "About Trout: The Antiquity of Southern Appalachian Brook Trout." *Trout,* Spring 1997, 47–48.

Behnke, R. J. "About Trout: Catchable Trout: Are Anglers Getting Their Money's Worth?" *Trout,* Winter 1996, 37–39.

Behnke, R. J. "About Trout: Coaster Brook Trout and Evolutionary Significance." *Trout*, Autumn 1994, 59–60.

Behnke, R. J. "About Trout: Do We Learn from History." *Trout*, Spring 1998, 55–57.

Behnke, R. J. "About Trout: Exceptions to the Rule." *Trout*, Fall 2006, 56–57.

Behnke, R. J. "About Trout: The First Forty Years: From Rhetoric to Research." *Trout*, Winter 1999, 52–54.

Behnke, R. J. "About Trout: Genetics: A Double-Edged Sword." *Trout*, Winter 2004, 59–60.

Behnke, R. J. "About Trout: Ghost Fish." *Trout*, Spring 2006, 56–57.

Behnke, R. J. "About Trout: Going Home Again: Revisiting Native Trout Watersheds of the West." *Trout*, Winter 1998, 55–56.

Behnke, R. J. "About Trout: Ivory-Billed Trout?" *Trout*, Fall 2005, 56–58.

Behnke, R. J. "About Trout: Limit Your Kill." *Trout*, Summer 1999, 54–55.

Behnke, R. J. "About Trout: The Nature and Nurture of Trout Behavior." *Trout*, Autumn 2002, 54–55.

Behnke, R. J. "About Trout: The Perils of Anadromy: What Have We Learned From History." *Trout*, Summer 1998, 57–59.

Behnke, R. J. "About Trout: Political Science." *Trout*, Fall 2004, 54–55.

Behnke, R. J. "About Trout: Prospecting for Native Trout." *Trout*, Spring 2005, 54–55.

Behnke, R. J. "About Trout: Science and Endangered Species." *Trout*, Winter 2008, 56–58.

Behnke, R. J. "About Trout: Tailwater Trout: Fish of Enormous Size." *Trout*, Spring 1996, 43–44.

Behnke, R. J. "About Trout: A Tale of Two Rivers." *Trout*, Autumn 1998, 65–67.

Behnke, R. J. "About Trout: Trading Stubbornness for Science." *Trout*, Fall 2007, 56–57.

Behnke, R. J. "About Trout: Trout Conservation at the Millennium." *Trout*, Spring 2000, 53–55.

Behnke, R. J. "About Trout: What Makes Salmon and Trout Run?" *Trout*, Summer 2002, 58–59.

Behnke, R. J. "About Trout: Wild Salmonid Genetics: An Impending Crisis." *Trout*, Summer 1995, 47–49.

Behnke, R. J. "About Trout: Wild Trout and Hatchery Trout: A 40-Year Review." *Trout*, Spring 1999, 59–60.

Behnke, R. J. "About Trout: Wild Trout, Hatchery Trout, and the Sad Story of Derisley Hobbs." *Trout*, Summer 2004, 59–60.

Behnke, R. J. "About Trout: Yellowstone Fishes: Changing Times and Changing Perspectives." *Trout*, Spring 1994, 55–59.

Behnke, R. J. "Comment: First Documented Case of Anadromy in a Population of Introduced Rainbow Trout in Patagonia, Argentina." *Transactions of the American Fisheries Society* 131 (2002): 582–585.

Behnke, R. J. "Livingston Stone, J. B. Campbell, and the Origins of Hatchery Rainbow Trout." *The American Fly Fisher* 16, no. 3 (1990): 20–22.

Behnke, R. J. *Native Trout of Western North America,* American Fisheries Society Monograph 6. Bethesda: American Fisheries Society, 1992.

Behnke, R. J. *Summary of Progress of Wild Trout Management, 1974–1989.* Paper presented at Wild Trout IV Symposium, 18–19 September 1989.

Behnke, R. J. *Trout and Salmon of North America.* New York: Free Press, 2002.

Benson, N. G. "The American Fisheries Society, 1920–1970." In *A Century of Fisheries in North America,* edited by N. G. Benson, 13–24. Washington: American Fisheries Society, 1970.

Binns, A., F. Eiserman, F. W. Jackson, A. F. Regenthal, and R. Stone. *The Planning, Operation, and Analysis of the Green River Fish Control Project.* Cheyenne and Salt Lake City: Utah State Department of Fish and Game and Wyoming Game and Fish Department, 1963.

Blaustein, A. R., D. G. Hokit, R. K. Ohara, and R. A. Holt. "Pathogenic Fungus Contributes to Amphibian Losses in the Pacific Northwest." *Biological Conservation* 67 (1994): 251–254.

Blinn, D. W., C. Runck, and D. A. Clark. "Effects of Rainbow Trout Predation on Little Colorado Spinedace." *Transactions of the American Fisheries Society* 22 (1993): 139–143.

Bogue, M. B. *Fishing the Great Lakes: An Environmental History, 1783–1933.* Madison: University of Wisconsin Press, 2000.

Born, S. M., and G. S. Stairs. *An Assessment of State Planning for Coldwater Fisheries Management in the United States.* Trout Unlimited, 2002, www .tu.org/atf/cf/%7BED0023C4-EA23-4396-9371-8509DC5B4953%7D/TU finalreport.pdf.

Bowen, J. T. "A History of Fish Culture As Related to the Development of Fishery Programs." In *A Century of Fisheries in North America,* edited by N. G. Benson, 71–93. Washington: American Fisheries Society, 1970.

Bowers, G. M. *U.S. Commission of Fish and Fisheries, Report of the Commissioner for the Year Ending June 30, 1899.* Washington: GPO, 1900.

Bowers, G. M. *U.S. Commission of Fish and Fisheries, Report of the Commissioner for the Year Ending June 30, 1903.* Washington: GPO, 1905.

Bowers, G. M. *U.S. Department of Commerce and Labor: Report of the Bureau of Fisheries, 1904.* Washington: GPO, 1905.

Bradford, D. F., F. Tabatabai, and D. M. Graber. "Isolation of Remaining Populations of the Native Frog, *Rana muscosa,* by Introduced Fishes in

Sequoia and Kings Canyon National Parks, California." *Conservation Biology* 7 (1993): 882–888.

British Museum. *Spreadsheet of Human Remains in the Collection of the British Museum.* The British Museum, www.thebritishmuseum.ac.uk/ pdf/human%20remains%20spreadsheet_.pdf (accessed 3 December 2008).

Brown, T. P., P. C. Rumsby, A. C. Capleton, L. Rushton, and L. S. Levy. "Pesticides and Parkinson's Disease—Is There a Link?" *Environmental Health Perspectives* 114 (2006): 156–164.

Brownlow, O. P. "Fish Planting in the High Sierra in 1929." *California Fish and Game* 16, no. 1 (1930): 8–12.

Brownlow, O. P. "Golden Trout Planting During 1928." *California Fish and Game* 15, no. 1 (1929): 23–28.

Bryant, H. C. "A Brief History of the California Fish and Game Commission." *California Fish and Game* (1921): 73–98.

Buchanan, J. M. *Public Finance in Democratic Process: Fiscal Institutions and Individual Choice.* Chapel Hill: University of North Carolina Press, 1987.

Bureau of Sport Fisheries and Wildlife. "Draft Statement, Proposed Program of Extension and Training." 16 January 1957. NACP RG22(266).

Bureau of Sport Fisheries and Wildlife. *Rare and Endangered Fish and Wildlife of the United States.* Bureau of Sport Fisheries and Wildlife Resource Publication 34. Washington: GPO, 1966.

Burrows, R. E. "Man-Made Fish." In *Sport Fishing USA*, edited by D. Saults, M. Walker, B. Hines, and R. G. Schmidt. Washington: GPO, 1971.

Busack, C. A., and K. P. Currens. "Genetic Risks and Hazards in Hatchery Operations: Fundamental Concepts and Issues." In *Uses and Effects of Cultured Fishes in Aquatic Ecosystems.* American Fisheries Society Symposium 15, edited by H. L. Schramm, Jr., and R. G. Piper, 71–80. Bethesda: American Fisheries Society, 1995.

Busack, C. A., and G. A. E. Gall. "Ancestry of Artificially Propagated California Rainbow Trout Strains." *California Fish and Game* 66, no. 1 (1980): 17–24.

Calabi, S. *Trout and Salmon of the World.* Edison, N.J.: Wellfleet, 1991.

Caldwell, M. *The Last Crusade: The War on Consumption, 1862–1954.* New York: Atheneum, 1988.

Caldwell, M. R., ed. *The Endangered Species Act: A Stanford Environmental Law Society Handbook.* Palo Alto: Stanford University Press, 2001.

California Fish and Game. "DFG Adopts Strategic Plan for Trout Management." California Department of Fish and Game, www.dfg.ca .gov/news/news04/04019.html (posted 18 February 2004, accessed 3 December 2008).

California Fish and Game. "Governor's Proposed Budget and Budget Bill: Budget Fact Book." California Department of Fish and Game, www.dfg.ca.gov/budget/08–09/08–09BudgetFactBook.pdf (accessed 8 December 2008).

California Fish and Game. *Report of the Commissioners of Fisheries of the State of California.* Years 1870–1892. Sacramento: Various publishers.

Cambray, J. A. "Impact on Indigenous Species Biodiversity Caused by the Globalisation of Alien Recreational Freshwater Fisheries." *Hydrobiologia* 500 (2003): 217–230.

Campton, D. E., and L. R. Kaeding. "Westslope Cutthroat Trout, Hybridization, and the U.S. Endangered Species Act." *Conservation Biology* 19, no. 4 (2005): 1323–1325.

Carson, R. L. *Silent Spring.* New York: Houghton Mifflin, 1962.

Cashner, F. M., G. R. Smith, and R. C. Cashner. "Robert Rush Miller and Frances Hubbs Miller." *Copeia* no. 4 (2003): 910–916.

Catholic World. "The Sanitary and Moral Condition of New York City." *Catholic World: A Monthly Magazine* 9, no. 53 (1869): 553–566. Catholic Publication House, available online at www.yale.edu/glc/archive/1021.htm.

Chandler, T. "Is the Pending World Record Rainbow a Real Trout?" Trout Underground, http://troutunderground.com/2007/06/27/is-the-pending-world-record-rainbow-a-real-trout (accessed 2 December 2008).

Cheney, A. N. "Angling Notes; Rainbow and Steelhead." *Forest and Stream,* 4 August 1894, xliii.

Clark, M. E., and K. A. Rose. "Factors Affecting Competitive Dominance of Rainbow Trout over Brook Trout in Southern Appalachian Streams: Implications of an Individual-Based Model." *Transactions of the American Fisheries Society* 126, no. 1 (1997): 1–20.

Cobb, E. W. "Pacific Salmon Fisheries." Appendix 13 in *Report of the United States Commissioner of Fisheries for the Fiscal Year 1930*, 409–704. Washington: GPO, 1931.

Cobb, E. W. "Results of Trout Tagging to Determine Migrations and Results from Plants Made." *Transactions of the American Fisheries Society* 63 (1933): 308–318.

Colorado Division of Wildlife. "New Strains of Rainbow Trout Hatch in Wild, Resistant to Whirling Disease." *DOW Insider,* in email message to author, 1 February 2008.

Colorado State Government. "Colorado State Archives Symbols and Emblems." Colorado Department of Personnel and Administration, www.colorado.gov/dpa/doit/archives/history/symbemb.htm#Fish (accessed 2 October 2008).

Colorado Wildlife Commission. "Wild and Gold Medal Trout Management." Colorado Wildlife Commission, rev. 12 June 2008.

Continental Congress. *The Association Agreed with the Grand American
Congress. October 20, 1774.* Available online through the Yale Law
School Avalon Project, http://avalon.law.yale.edu/18th_century/
contcong_10-20-74.asp (accessed 3 December 2008).

Cooke, S. J., and I. G. Cowx. "The Role of Recreational Fishing in Global
Fish Crises." *Bioscience* 54, no. 9 (2004): 857–859.

Cooper, E. L. "Returns from Plantings of Legal-Sized Brook, Brown and
Rainbow Trout in the Pigeon River, Otsego County, Michigan." *Transactions of the American Fisheries Society* (1956): 265–280.

Cordes, J. F., M. R. Stephens, M. A. Blumberg, and B. May. "Identifying Introgressive Hybridization in Native Populations of California
Golden Trout Based on Molecular Markers." *Transactions of the American Fisheries Society* 135, no. 1 (2006): 110–128.

Court of Assistants of the Colony of the Massachusetts Bay. *Records of
the Court of Assistants of the Colony of the Massachusetts Bay, 1630–1692.*
Boston: County of Suffolk, 1904.

Courtenay, W. R. "The Case for Caution with Fish Introductions." In *Uses
and Effects of Cultured Fishes in Aquatic Ecosystems,* American Fisheries
Society Symposium 15, edited by H. L. Schramm, Jr., and R. G. Piper,
413–424. Bethesda: American Fisheries Society, 1995.

Courtenay, W. R., Jr., and J. E. Deacon. "Fish Introductions in the American Southwest—A Case-History of Rogers Spring, Nevada." *Southwestern Naturalist* 28, no. 2 (1983): 221–224.

Courtenay, W. R., Jr., D. A. Hensley, J. N. Taylor, and J. A. McCann.
"Distribution of Exotic Fishes in the Continental United States." In
Distribution, Biology, and Management of Exotic Fishes, edited by W. R.
Courtenay, Jr., and J. R. Stauffer, Jr., 41–77. Baltimore: Johns Hopkins
University Press, 1984.

Courtenay, W. R., Jr., and J. R. Stauffer. "Introduction." In *Distribution,
Biology, and Management of Exotic Fishes,* edited by W. R. Courtenay,
Jr., and J. R. Stauffer, Jr., xiii–xiv. Baltimore: Johns Hopkins University
Press, 1984.

Cronon, W. *Changes in the Land: Indians, Colonists, and the Ecology of New
England.* New York: Hill and Wang, 1983.

Cronon, W. "Introduction: In Search of Nature." In *Uncommon Ground:
Rethinking the Human Place in Nature,* edited by W. Cronon, 23–56.
New York: Norton, 1996.

Cronon, W. "The Trouble with Wilderness; or, Getting Back to the Wrong
Nature." In *Uncommon Ground: Rethinking the Human Place in Nature,*
edited by W. Cronon, 69–90. New York: Norton, 1996.

Currens, K. P., A. R. Hemmingsen, R. A. French, D. V. Buchanan, C. B. Schrek, and H. W. Li. "Introgression and Susceptibility to Disease in a Wild Population of Rainbow Trout." *North American Journal of Fisheries Management* 17 (1997): 1065–1078.

Cutter, R. "Gorgeous Goldens." *Trout,* Autumn 2001: 16–62.

de C. Ravenal, W. "Report on the Propagation and Distribution of Food-Fishes." In *U.S. Commission of Fish and Fisheries, Report of the Commissioner for the Year Ending June 30, 1899,* xxxv–cxviii. Washington: GPO, 1900.

Denton, J. A. *Rocky Mountain Radical: Myron W. Reed, Christian Socialist.* Albuquerque: University of New Mexico Press, 1997.

Docker, M. F., A. Dale, and D. D. Heath. "Erosion of Interspecific Reproductive Barriers Resulting from Hatchery Supplementation of Rainbow Trout Sympatric with Cutthroat Trout." *Molecular Ecology* 12 (2003): 3515–3521.

Dollar, A. M., and M. Katz. "Rainbow Trout Brood Stocks and Strains in American Hatcheries as Factors in the Occurrence of Hepatoma." *The Progressive Fish-Culturist* 26 (1964): 167–174.

Doyle, R. W., C. Herbinger, C. T. Taggart, and S. Lochmann. "Use of DNA Microsatellite Polymorphism to Analyze Genetic Correlations Between Hatchery and Natural Fitness." In *Uses and Effects of Cultured Fishes in Aquatic Ecosystems,* American Fisheries Society Symposium 15, edited by H. L. Schramm, Jr., and R. G. Piper, 205–211. Bethesda: American Fisheries Society, 1995.

Drake, D. C., and R. J. Naiman. "An Evaluation of Restoration Efforts in Fishless Lakes Stocked with Exotic Trout." *Conservation Biology* 14, no. 6 (2000): 1807–1820.

Dulles, F. R. *A History of Recreation.* New York: Appleton-Century-Crofts, 1965.

Duncan, D. *Miles from Nowhere: Tales from America's Contemporary Frontier.* Lincoln: University of Nebraska Press, 2000.

Dunham, J. B., S. B. Adams, R. E. Schroeter, and D. C. Novinger. "Alien Invasions in Aquatic Ecosystems: Toward an Understanding of Brook Trout Invasions and Potential Impacts on Inland Cutthroat Trout in Western North America." *Reviews in Fish Biology and Fisheries* 12, no. 4 (2002): 373–391.

Dunham, J. B., D. S. Pilliod, and M. K. Young. "Assessing the Consequences of Nonnative Trout in Headwater Ecosystems in Western North America." *Fisheries* 29, no. 6 (2004): 18.

Dunlap, T. R. "Remaking the Land: The Acclimatization Movement and Anglo Ideas of Nature." *Journal of World History* 8, no. 2 (1997): 303–319.

Dunlap, T. R. "Sport Hunting and Conservation, 1880–1920." *Environmental Review* (1988): 51–60.

El-Matbouli, M., R. Hoffmann, and M. P. Küppers. "Identification of a Whirling Disease Resistant Strain of Rainbow Trout in Germany." Whirling Disease Initiative, http://whirlingdisease.montana.edu/biblio/pdfs/WDProceedings_2002.pdf (accessed 7 December 2008).

Epifanio, J. "The Status of Coldwater Fishery Management in the United States: An Overview of State Programs." *Fisheries* 25, no. 7 (2000): 13–27.

Epifanio, J., and W. Fosburgh. *A Status Report of Coldwater Fishery Management in the U.S.: An Overview of State Programs.* Trout Unlimited, 1998, www.tu.org/pdf/newsstand/library/statemgmt.pdf.

Epifanio, J., and D. Nickum. *Fishing for Answers: Status and Trends for Coldwater Fisheries Management in Colorado.* Trout Unlimited, 1997, www.cotrout.org/Portals/0/pdf/reports/EpifanioNickum97a.pdf.

Epifanio, J., and J. Nielsen. "The Role of Hybridization in the Distribution, Conservation and Management of Aquatic Species." *Reviews in Fish Biology and Fisheries* 10 (2000): 245–251.

Evermann, B. W., and H. C. Bryant. "California Trout." *California Fish and Game* 5, no. 3 (1919): 105–135.

Ewen, S. *Captains of Consciousness: Advertising and the Social Roots of the Consumer Culture.* New York: McGraw-Hill, 1976.

Faist, C. *State of California Department of Fish and Game, Air Operations, 1945–1983.* (Newspaper articles and reminiscences, compiled by Faist in 2006.) Available from the California Department of Fish and Game.

Farquhar, F. P. *Up and Down California in 1860–1864: The Journal of William H. Brewer, Professor of Agriculture in the Sheffield Scientific School from 1864–1903.* Reprint, Berkeley: University of California Press, 1966.

Fearnow, E. C. *Stocking Interior Waters of the United States: Bureau of Fisheries Document No. 941a.* Washington: GPO, 1924.

Fedler, A. J., and R. B. Ditton. "Dropping Out and Dropping In: A Study of Factors for Changing Recreational Fishing Participation." *North American Journal of Fisheries Management* 21, no. 2 (2001): 283–292.

Fedler, A. J., and R. B. Ditton. "Understanding Angler Motivations in Fisheries Management." *Fisheries* 19, no. 4 (1994): 6–13.

Fenner, D. B., M. G. Walsh, and D. L. Winkelman. "Diet Overlap of Introduced Rainbow Trout and Three Native Fishes in an Ozark Stream." In *Propagated Fish in Resource Management,* American Fisheries Society, Symposium 44, edited by M. J. Nickum, P. M. Mazik, J. G. Nickum, and D. D. MacKinlay, 475–482. Bethesda: American Fisheries Society, 2004.

Firestone, J., and R. Barber. "Fish as Pollutants: Limitations of and Cross-currents in Law, Science, Management, and Policy." *Washington Law Review* 78, no. 3 (2003): 693–756.

Fisher, W. L., and J. P. Burroughs. "Stream Fisheries Management in the United States." *Fisheries* 28, no. 2 (2003): 10–18.

Forbes, S. A. "The Investigation of a River System in the Interest of Its Fisheries." *Transactions of the American Fisheries Society* 40 (1911): 179–193.

Forest and Stream. "American Acclimatization Society." 22 November 1877.

Forest and Stream. "Defiance of Law in New York." 25 August 1881.

Forest and Stream. "Fish Culture in New York." 3 August 1876.

Forest and Stream. "Forest Park Associations and Proprietary Clubs." 22 June 1876.

Forest and Stream. "Game Protection." 25 October 1877.

Forest and Stream. "New Game Preserve on Long Island." 26 November 1874.

Forest and Stream. "Rainbow Trout in England." 7 May 1898.

Forest and Stream. "Trout Day." 29 March 1883.

Forest and Stream. "Trout Poaching on Long Island." 15 July 1875.

Forest and Stream. "United States Fish Hatching in California." 17 September 1874.

Forest and Stream. "Vermont Game Laws." 14 December 1876.

Forest and Stream. "Wonderful Supply of Rainbow Trout in Michigan." 25 July 1914.

Fox, S. *The American Conservation Movement: John Muir and His Legacy.* Madison: University of Wisconsin Press, 1981.

Fradkin, P. L. *A River No More: The Colorado River and the West.* Berkeley: University of California Press, 1996.

Freud, S. *Civilization and Its Discontents.* Translated and edited by James Strachey. New York: W. W. Norton. 2005.

Fuller, P. L., L. G. Nico, and J. D. Williams. *Nonindigenous Fishes Introduced into Inland Waters of the United States.* Bethesda: American Fisheries Society, Special Publication 27, 1999.

Gabelhouse, D. W., Jr. "Staffing, Spending, and Funding of State Inland Fisheries Programs." *Fisheries* 30 (2005): 10–17.

Gates, J. M. "Economics of Recreational Fishing Collapse." *Fisheries* 27, no. 3 (2002): 34–35.

Gay, J. "Rainbow Trout Planting." *Forest and Stream,* 1 October 1891.

Giles, R. A. *Shasta County, California: A History.* Oakland, Calif.: Biobooks, 1949.

Goode, G. B. "The Status of the U.S. Fish Commission in 1884." In *United*

States Commission of Fish and Fisheries, Report of the Commissioner for 1884, 1139–1180. Washington: GPO, 1886.

Gordon, D., D. W. Chapman, and T. C. Bjornn. "Economic Evaluation of Sport Fisheries." *Transactions of the American Fisheries Society* 102, no. 2 (1973): 293–311.

Gordon, S. "Now Trout Fly Too!" *Sports Afield,* August 1951.

Gordon, T. "Rainbow Trout." *Forest and Stream,* 13 July 1907.

Gottlieb, R. *Forcing the Spring: The Transformation of the American Environmental Movement.* Washington, D.C.: Island, 1993.

Green, S. "Letter from Seth Green." *Forest and Stream,* 8 August 1878.

Green, S. "Letter from Seth Green." *Forest and Stream,* 10 October 1878.

Green, S. "The Rainbow Trout." *American Angler,* 13 March 1886.

Green, S. "Rearing of California Mountain Trout (*Salmo irideus*)." *Bulletin of the United States Fish Commission for 1881,* 23. Washington: GPO, 1882.

Green, S. "Seth Green and Fish Culture." Undated letter to unnamed correspondent in the Rochester Museum and Science Center, Rochester, N.Y.

Green, S. "Seth Green Paper on Stocking Waters with Fish." *Transactions of the American Fisheries Society* 8 (1879): 22–27.

Greer, L. S. "The United States Forest Service and the Postwar Commodification of Outdoor Recreation." In *For Fun and Profit: The Transformation of Leisure into Consumption,* edited by R. Butsch, 152–170. Philadelphia: Temple University Press, 1990.

Griffith, G. A. *For the Love of Trout.* Grayling, Mich.: The George Griffith Foundation, 1993.

Grinnell, J., and T. I. Storer. *Animal Life in the Yosemite.* Berkeley: University of California Press, 1924.

Gross, A. S. *The Changing Shape of the American Landscape: Travel, Corporate Expansion, and Consumer Culture, 1845–1945.* Ph.D. diss. University of California, Davis, 2001.

Grubic, G., and A. Herd. "Astraeus Origins." *The American Fly Fisher* 27, no. 4 (Fall 2001): 16–22.

Guinet, L. "Remy et Gehin, les Inventeurs Français de la Pisciculture." *Isis* 5, no. 2 (1923): 403–405.

Hagen, W., and J. P. O'Connor. "Public Fish Culture in the United States, 1958: A Statistical Summary." Washington: GPO, 1959.

Hallock, C. *The Fishing Tourist.* New York: Harper and Brothers, 1873.

Halverson, M. A. "Stocking Trends: A Quantitative Review of Governmental Fish Stocking in the United States, 1931 to 2004." *Fisheries* 33, no. 2 (2008): 69–75.

Harris, L. E., "Evolution of Colorado's Stocking Procedures for Fish Originating from *Myxobolus cerebralis* (Whirling Disease) Positive

Hatcheries." In-house report. Denver: Colorado Division of Wildlife, 1995.

Hauer, F. R., J. A. Stanford, and M. S. Lorang. "Pattern and Process in Northern Rocky Mountain Headwaters: Ecological Linkages in the Headwaters of the Crown of the Continent." *Journal of the American Water Resources Association* 43, no. 1 (2007): 104–117.

Hays, S. P. *Conservation and the Gospel of Efficiency*. Pittsburgh: University of Pittsburgh Press, 1959.

Hazzard, A. S., and D. S. Shetter. "Results from Experimental Plantings of Legal-Sized Brook Trout (*Salvelinus fontinalis*) and Rainbow Trout (*Salmo irideus*)." *Transactions of the American Fisheries Society* 68 (1939): 196–210.

Hearn, W. E., "Interspecific Competition and Habitat Segregation Among Stream-Dwelling Trout and Salmon: A Review." *Fisheries* 12, no. 5 (1987): 24–31.

Hedgpeth, J. W. "Livingston Stone and Fish Culture in California." *California Fish and Game* 27, no. 8 (1941).

Heidinger, R. C. "Stocking for Sport Fisheries Enhancement." In *Inland Fisheries Management in North America*, edited by C. C. Kohler and W. A. Hubert, 309–333. Bethesda: American Fisheries Society, 1993.

Henshall, J. A. "Black Bass in Trout Waters." *Forest and Stream*, 23 July 1886.

Hilborn, R. "Hatcheries and the Future of Salmon in the Northwest." *Fisheries* 17, no. 1 (1992): 5–8.

Hitt, N. P., C. A. Frissell, C. C. Muhlfeld, and F. W. Allendorf. "Spread of Hybridization Between Native Westslope Cutthroat Trout, *Oncorhynchus clarki lewisi*, and Nonnative Rainbow Trout, *Oncorhynchus mykiss*." *Canadian Journal of Fisheries and Aquatic Sciences* 60, no. 12 (2003): 1440–1451.

Holden, P. B. "Ghosts of the Green River: Impacts of Green River Poisoning on Management of Native Fishes." In *Battle Against Extinction: Native Fish Management in the American West*, edited by W. L. Minckley and J. E. Deacon, 43–54. Tucson: University of Arizona Press, 1991.

Holmes, O. W. "The Autocrat of the Breakfast-Table: Every Man His Own Boswell." *Atlantic Monthly*, May 1858: 871–883, available online through the Cornell University Library at http://cdl.library.cornell.edu/moa/.

Hoover, E. E., and M. S. Johnson. "Migration and Depletion of Stocked Brook Trout." *Transactions of the American Fisheries Society* 67 (1938): 224–227.

Horak, D. "Native and Nonnative Fish Species Used in State Fisheries Management Programs in the United States." In *Uses and Effects of Cultured Fishes in Aquatic Ecosystems*, American Fisheries Society Symposium 15, edited by H. L. Schramm, Jr., and R. G. Piper, 61–67. Bethesda: American Fisheries Society, 1995.

Hummel, R. *Hunting and Fishing for Sport: Commerce, Controversy, Popular Culture.* Bowling Green: Bowling Green State University Popular Press, 1994.

Hyatt, W. A., M. Humphreys, and N. T. Hagstrom. *A Trout Management Plan for Connecticut's Rivers and Streams.* Hartford: Connecticut Department of Environmental Protection, 1999.

Jackson, J. R., J. C. Boxrucker, and D. W. Willis. "Trends in Agency Use of Propagated Fishes as a Management Tool in Inland Fisheries." In *Propagated Fish in Resource Management,* American Fisheries Society Symposium 44, edited by M. J. Nickum, P. M. Mazik, J. G. Nickum, and D. D. MacKinlay, 131–138. Bethesda: American Fisheries Society, 2004.

Jelks, H. L., S. J. Walsh, N. M. Burkhead, S. Contreras-Balderas, E. Díaz-Pardo, D. A. Hendrickson, J. Lyons, N. E. Mandrak, F. McCormick, J. S. Nelson, S. P. Platania, B. A. Porter, C. B. Renaud, J. J. Schmitter-Soto, E. B. Taylor, and M. L. Warren, Jr. "Conservation Status of Imperiled North American Freshwater and Diadromous Fishes." *Fisheries* 33, no. 8 (2008): 372–407.

Johnson, D. M., R. J. Behnke, D. A. Harpman, and R. G. Walsh. "Economic Benefits and Costs of Stocking Catchable Rainbow Trout: A Synthesis of Economic Analysis in Colorado." *North American Journal of Fisheries Management* 15 (1995): 26–32.

Jordan, D. S., and B. W. Evermann. *American Food and Game Fishes.* New York: Doubleday, Page, 1905.

Kamel, F., C. M. Tanner, D. M. Umbach, J. A. Hoppin, M. C. R. Alavanja, A. Blair, K. Comyns, S. M. Goldman, M. Korell, J. W. Langston, G. W. Ross, and D. P. Sandier. "Pesticide Exposure and Self-reported Parkinson's Disease in the Agricultural Health Study." *American Journal of Epidemiology* 165 (2007): 364–374.

Kerans, B. L., and A. V. Zale. "The Ecology of *Myxobolus cerebralis.*" In *Whirling Disease: Reviews and Current Topics,* American Fisheries Society Symposium 29, edited by J. L. Bartholomew and J. C. Wilson, 145–166. Bethesda: American Fisheries Society, 2002.

Kincaid, H. L. "An Evaluation of Inbreeding and Effective Population Size in Salmonid Broodstocks in Federal and State Hatcheries." In *Uses and Effects of Cultured Fishes in Aquatic Ecosystems,* American Fisheries Society Symposium 15, edited by H. L. Schramm, Jr., and R. G. Piper, 193–204. Bethesda: American Fisheries Society, 1995.

Kincaid, H. L., M. J. Gray, L. J. Mengel, and S. Brimm. *National Fish Strain Registry—Trout; Species Tables of Reported Strains and Broodstocks.* Wellsboro, Pa.: U.S. Geological Survey, 1997.

Kinsey, D. "'Seeding the Water as the Earth': The Epicenter and Peripheries of a Western Aquacultural Revolution." *Environmental History* 11, no. 3 (2006): 527–566.

Kittams, W. H. *Rotenone Treatment of Green River as Related to Dinosaur National Monument.* National Park Service Report, UMMZ.

Klinko, D. W. *Ante-bellum American Sporting Magazines and the Development of a Sportsmen's Ethic.* Ph.D. diss. Washington State University, 1986.

Knapp, R. A. "Non-Native Trout in Natural Lakes of the Sierra Nevada: An Analysis of Their Distribution and Impacts on Native Aquatic Biota." In *Sierra Nevada Ecosystem Project: Final Report to Congress, vol. III, Assessments and Scientific Basis for Management Options.* Davis: University of California, Centers for Water and Wildland Resources, 1996.

Knapp, R. A. "Presentation at American Museum of Natural History Freshwater Biodiversity Conference," 8 April 2005.

Knapp, R. A., P. S. Corn, and D. E. Schindler. "The Introduction of Nonnative Fish into Wilderness Lakes: Good Intentions, Conflicting Mandates, and Unintended Consequences." *Ecosystems* 4 (2000): 275–278.

Knapp, R. A., and K. R. Matthews. "Non-Native Fish Introductions and the Decline of the Mountain Yellow-Legged Frog from Within Protected Areas." *Conservation Biology* 14, no. 2 (2000): 428–438.

Knapp, R. A., R. A. Matthews, and O. Sarnelle. "Resistance and Resilience of Alpine Lake Fauna to Fish Introductions." *Ecological Monographs* 71, no. 3 (2001): 401–421.

Knowles, J. *Alone in the Wilderness.* Boston: Small, Maynard, 1913.

Knox, T. W. "Summer Clubs on Great South Bay." *Harper's New Monthly Magazine* 61, no. 362 (1880): 206–219.

Kolar, C. S., and D. M. Lodge. "Ecological Predictions and Risk Assessment for Alien Fishes in North America." *Science* 298, no. 5596 (2002): 1233–1236.

Konrad, A., and S. Konrad. "Rainbow Trout World Record Hunt." Trophy Trout Guide, www.trophytroutguide.com/articles/world_record_rainbow_trout_adam.htm (accessed 3 December 2008).

Kozfkay, J. R., D. J. Schill, and D. M. Teuscher. "Improving Vulnerability to Angling of Rainbow Trout." In *Propagated Fish in Resource Management,* American Fisheries Society Symposium 44, edited by M. J. Nickum, P. M. Mazik, J. G. Nickum, and D. D. MacKinlay, 497–504. Bethesda: American Fisheries Society, 2004.

Krukoff, B. A., and A. C. Smith. "Rotenone Yielding Plants of South America." *American Journal of Botany* 24, no. 9 (1937): 573–587.

Kruse, C. G., W. A. Hubert, and F. J. Rahel. "Status of Yellowstone Cutthroat Trout in Wyoming Waters." *North American Journal of Fisheries Management* 20 (2000): 693–705.

La Motte, A. "Fish Propagation in California: A Study of Practical Trout Raising." *Overland Monthly* 32 (1898): 237–239.

Landres, P., S. Meyer, and S. Matthews. "The Wilderness Act and Fish Stocking: An Overview of Legislation, Judicial Interpretation, and Agency Implementation." *Ecosystems* 4, no. 4 (2001): 287–295.

Larson, G. L., S. E. Moore, and D. C. Lee. "Angling and Electrofishing for Removing Nonnative Rainbow Trout from a Stream in a National Park." *North American Journal of Fisheries Management* 6 (1986): 580–585.

Lassuy, D. R. "Introduced Species as a Factor in Extinction and Endangerment of Native Fish Species." In *Uses and Effects of Cultured Fishes in Aquatic Ecosystems*, American Fisheries Society Symposium 15, edited by H. L. Schramm, Jr., and R. G. Piper, 391–396. Bethesda: American Fisheries Society, 1995.

Leach, G. C. "Propagation and Distribution of Food Fishes, Fiscal Year 1931." In *Report of the United States Commissioner of Fisheries for the Fiscal Year 1931*. Washington: GPO, 1932.

Leach, G. C., M. C. James, and E. J. Douglass. "Propagation and Distribution of Food Fishes, Fiscal Year 1939." In *Report of the United States Commissioner of Fisheries for the Fiscal Year 1939*, 555–598. Washington: GPO, 1941.

Lears, T. J. J. *No Place of Grace: Antimodernism and the Transformation of American Culture.* Chicago: University of Chicago Press, 1981.

Leary, R. F., F. W. Allendorf, and G. K. Sage. "Hybridization and Introgression Between Introduced and Native Fish." In *Uses and Effects of Cultured Fishes in Aquatic Ecosystems*, American Fisheries Society Symposium 15, edited by H. L. Schramm, Jr., and R. G. Piper, 91–101. Bethesda: American Fisheries Society, 1995.

Lee, D. P. "Contribution of Nonnative Fish to California's Inland Recreational Fishery." In *Uses and Effects of Cultured Fishes in Aquatic Ecosystems*, American Fisheries Society Symposium 15, edited by H. L. Schramm, Jr., and R. G. Piper, 16–20. Bethesda: American Fisheries Society, 1995.

Lee, R. M., and J. N. Rinne. "Critical Thermal Maxima of Five Trout Species in the Southwestern United States." *Transactions of the American Fisheries Society* 109 (1980): 632–635.

Leitritz, E. *A History of California's Fish Hatcheries, 1870–1960*, Fish Bulletin 150. Sacramento: California Department of Fish and Game, 1970.

Lennon, R. E., J. B. Hunn, R. A. Schnick, and R. E. Buress. "Reclamation of Ponds, Lakes, and Streams with Fish Toxicants: A Review." Food and Agriculture Organization, 1971, www.fao.org/docrep/003/b0465e/b0465e06.htm (accessed 3 December 2008).

Leonard, J. R. *The Fish Car Era of the National Fish Hatchery System.* Washington: GPO, 1979.

Leopold, A. *Game Management.* Madison: University of Wisconsin Press, 1987.

Leopold, A. S., S. A. Cain, C. M. Cottam, I. N. Gabrielson, and T. L. Kimball. "Wildlife Management in the National Parks." *Transactions of the North American Wildlife and Natural Resources Conference* 28 (1963): 29–44.

Lever, C. *Naturalized Birds of the World.* New York: John Wiley and Sons, 1987.

Lewis, M., and W. Clark. *The Journals of the Lewis and Clark Expedition.* Available online at the University of Nebraska-Lincoln Libraries-Electronic Text Center, http://lewisandclarkjournals.unl.edu.

Lewis, R. C. "Selective Breeding of Rainbow Trout at Hot Creek Hatchery." *California Fish and Game* 30, no. 1 (1944): 95–97.

Li, H. W., and P. B. Moyle. "Management of Introduced Fishes." In *Inland Fisheries Management in North America,* edited by C. C. Kohler and W. A. Hubert, 287–307. Bethesda: American Fisheries Society, 1993.

Lichatowich, J. *Salmon Without Rivers: A History of the Pacific Salmon Crisis.* Washington, D.C.: Island, 1999.

Liknes, G. A., and P. J. Graham. "Westslope Cutthroat in Montana: Life History, Status, and Management." In *Status and Management of Interior Stocks of Cutthroat Trout,* American Fisheries Society Symposium 4, edited by R. E. Gresswell, 53–60. Bethesda: American Fisheries Society, 1988.

Lubowski, R. N., M. Vesterby, S. Bucholtz, A. Baez, and M. J. Roberts. *Major Uses of Land in the United States, 2002.* U.S. Department of Agriculture Economic Research Service, Economic Information Bulletin 14, May 2006. www.ers.usda.gov/publications/EIB14/eib14.pdf (accessed 13 December 2008).

Lucas, M. D., R. E. Drew, P. A. Wheeler, P. A. Verrell, and G. A. Thorgaard. "Behavioral Differences Among Rainbow Trout Clonal Lines." *Behavior Genetics* 34, no. 3 (2004): 355–365.

McCann, J. A. "Preface." In *Distribution, Biology, and Management of Exotic Fishes,* edited by W. R. Courtenay, Jr., and J. R. Stauffer, Jr., ix–x. Baltimore: Johns Hopkins University Press, 1984.

McCann, J. A. "Involvement of the American Fisheries Society with Exotic Species, 1969–1982." In *Distribution, Biology, and Management of Exotic Fishes,* edited by W. R. Courtenay, Jr., and J. R. Stauffer, Jr., 1–7. Baltimore: Johns Hopkins University Press, 1984.

McCaull, J. "Dams of Pork." *Environment* 17, no. 1 (January 1975): 11–16.

MacCrimmon, H. R. "World Distribution of Rainbow Trout (*Salmo gaird-*

neri)." *Journal of the Fisheries Research Board of Canada* 28, no. 5 (1971): 663–704.

McEvoy, A. F. *The Fisherman's Problem: Ecology and Law in the California Fisheries, 1850–1980.* New York: Cambridge University Press, 1986.

McEvoy, A. F. "Science, Culture, and Politics in U.S. Natural Resouces Management." *Journal of the History of Biology* 25, no. 3 (1992): 469–486.

McGurrin, J., C. Ubert, and D. Duff. "Use of Cultured Salmonids in the Federal Aid in Sport Fish Restoration Program." In *Uses and Effects of Cultured Fishes in Aquatic Ecosystems,* American Fisheries Society Symposium 15, edited by H. L. Schramm, Jr., and R. G. Piper, 12–15. Bethesda: American Fisheries Society, 1995.

McHugh, J. L. "Trends in Fishery Research." In *A Century of Fisheries in North America,* edited by N. G. Benson, 25–56. Washington: American Fisheries Society, 1970.

Madsen, D. H. "Protection of Native Fishes in the National Parks." *Transactions of the American Fisheries Society* 66, no. 1 (1937): 395–397.

Magoulick, D. D., and M. A. Wilzbach. "Effect of Temperature and Macrohabitat on Interspecific Aggression, Foraging Success, and Growth of Brook Trout and Rainbow Trout Pairs in Laboratory Streams." *Transactions of the American Fisheries Society* 127 (1998): 708–717.

Marchetti, M. P., and G. A. Nevitt. "Effects of Hatchery Rearing on Brain Structures of Rainbow Trout, *Oncorhynchus mykiss*." *Environmental Biology of Fishes* 66 (1963): 9–14.

Marsh, G. P. "Report, Made Under Authority of the Legislature of Vermont, on the Artificial Propagation of Fish." 1857. Reprinted in *So Great a Vision: The Conservation Writings of George Perkins Marsh,* edited by Stephen C. Trombulak. Hanover, N.H.: Middlebury College Press, 2001.

Martin, J., J. Webster, and G. Edwards. "Hatcheries and Wild Stocks: Are They Compatible." *Fisheries* 17, no. 1 (1992): 4.

Martinez, A. M. "Identification and Status of Colorado River Cutthroat Trout in Colorado." In *Status and Management of Interior Stocks of Cutthroat Trout,* American Fisheries Society Symposium 4, edited by R. E. Greswell, 81–89. Bethesda: American Fisheries Society, 1988.

Marx, L. "Environmental Degradation and the Ambiguous Social Role of Science and Technology." *Journal of the History of Biology* 25, no. 3 (1992): 449–468.

Mather, F. "Progress in Fish Culture." *Century Illustrated Magazine* 27, no. 6 (1884).

Matthews, K. R., K. L. Pope, H. K. Preisler, and R. A. Knapp. "Effects of Nonnative Trout on Pacific Treefrogs (*Hyla regilla*) in the Sierra Nevada." *Copeia,* no. 4 (2001): 1130–1137.

Mayr, E. *The Growth of Biological Thought.* Cambridge: Belknap Harvard, 1982.

Metcalf, J. L., V. L. Pritchard, S. M. Silvestri, J. B. Jenkins, J. S. Wood, D. E. Cowley, R. P. Evans, D. K. Shiozawa, and A. P. Martin. "Across the Great Divide: Genetic Forensics Reveals Misidentification of Endangered Cutthroat Trout Populations." *Molecular Ecology* 16, no. 21 (2007): 4445–4454.

Meyer, F. P. "Aquaculture Disease and Health Management." *Journal of Animal Science* 69 (1991): 4201–4208.

Miller, L. M., T. Close, and A. R. Kapuscinski. "Lower Fitness of Hatchery and Hybrid Rainbow Trout Compared to Naturalized Populations in Lake Superior Tributaries." *Molecular Ecology* 13 (2004): 3379–3388.

Miller, M. P., and E. R. Vincent. "Rapid Natural Selection for Whirling Disease Resistance in Rainbow Trout from Harrison Lake, Montana." Whirling Disease Initiative, http://whirlingdisease.montana.edu/biblio/pdfs/Symposia/Miller_WDS2007.pdf (accessed 8 December 2008).

Miller, R. R. "Is Our Native Underwater Life Worth Saving?" *National Parks Magazine*, 1 May 1963.

Miller, R. R., and E. P. Pister. "Management of the Owens Pupfish, *Cyprinodon radiosus,* in Mono County, California." *Transactions of the American Fisheries Society* 100, no. 3 (1971): 502–509.

Miller, R. R., J. D. Williams, and J. E. Williams. "Extinctions of North American Fishes During the Past Century." *Fisheries* 14, no. 6 (1989): 22–38.

Milliron, C., P. L. Kiddoo, M. Lockhart, and R. Ziegler. *Aquatic Biodiversity Management Plan for the West Walker Basin of the Sierra Nevada, Mono County, California, 2004–2014.* Sacramento: California Department of Fish and Game.

Minckley, W. L., and M. E. Douglas. "Discovery and Extinction of Western Fishes: A Blink in the Eye in Geologic Time." In *Battle Against Extinction: Native Fish Management in the American West,* edited by W. L. Minckley and J. E. Deacon, 7–18. Tucson: University of Arizona Press, 1991.

Modin, J. "Whirling Disease in California: A Review of Its History, Distribution, and Impacts, 1965–1997." *Journal of Aquatic Animal Health* 10 (1998): 132–142.

Montana Outdoors. "Why Montana Went Wild." 1 November 2004.

Moore, S. E., M. A. Kulp, J. Hammonds, and B. Rosenlund. "Restoration of Sams Creek and an Assessment of Brook Trout Restoration Methods: Great Smoky Mountains National Park." U.S. National Park Service Technical Report NPS/NRWRD/NRTR-2005/342.

Moyle, P. B. "Fish Introductions into North America: Patterns and Ecological Impact." In *Ecology of Biological Invasions of North America and Hawaii*, edited by H. A. Mooney and J. A. Drake, 27–43. New York: Springer-Verlag, 1986.

Moyle, P. B. "The Importance of an Historical Perspective: Fish Introductions." *Fisheries* 22, no. 10 (1997): 14.

Moyle, P. B., P. K. Crain, K. Whitener, and J. F. Mount. "Alien Fishes in Natural Streams: Fish Distribution, Assemblage Structure, and Conservation in the Cosumnes River, California, U.S.A." *Environmental Biology of Fishes* 68 (2003): 143–162.

Muhlfeld, C. C., S. T. Kalinowski, T. E. McMahon, M. L. Taper, S. Painter, R. F. Leary, F. W. Allendorf. "Hybridization Rapidly Reduces Fitness of a Native Trout in the Wild." *Biology Letters*, 5, no. 3 (2009): 328–331.

Muir, J. *The Mountains of California*. New York: Century, 1894.

Muir, J. *Our National Parks*. New York: Houghton, Mifflin, 1901.

Muir, J. "Salmon-Breeding." *San Francisco Bulletin*, 29 October 1874.

Mumford, L. *The City in History*. New York: Harcourt, Brace and World, 1961.

Munday, P. "Rivers of Our Discontent: Montana Puts Limits on National Trout Unlimited." *High Country News*, 16 April 2007.

Nash, R. F. *Wilderness and the American Mind*. 4th ed. New Haven: Yale University Press, 2001.

National Agricultural Statistics Service. *Trout Production*. National Agricultural Statistics Service, http://usda.mannlib.cornell.edu/usda/nass/TrouProd//2000s/2006/TrouProd-02–24–2006.pdf (accessed 24 February 2006).

National Fish Hatchery Review Panel. *Report of the National Fish Hatchery Review Panel*. Arlington: The Conservation Fund, 1994.

Needham, P. R., and R. J. Behnke. "The Origin of Hatchery Rainbow Trout." *The Progressive Fish Culturist* 24, no. 4 (1962): 156–158.

Nehring, R. B. "Biological, Environmental, and Epidemiological Evidence Implicating *Myxobolus cerebralis* (the Protozoan Parasite Causing Whirling Disease) in the Dramatic Decline of the Wild Rainbow Trout Population in the Colorado River in Middle Park, Colorado." Denver: Colorado Division of Wildlife, 1993.

Nehring, R. B. "Stream Fisheries Investigations: Federal Aid Project F-237-R6." Denver: Colorado Division of Wildlife, 1999.

Nehring, R. B. "Stream Fisheries Investigations: Federal Aid Project F-237-R11." Denver: Colorado Division of Wildlife, 2004.

New York Commissioners of Fisheries. *Report of the Commissioners of Fisheries*. Years 1869–1885. Albany: various publishers.

Nichols, T. L. *Forty Years of American Life*. London: Longmans, Green, 1874.

Nickum, D. "Whirling Disease in the United States: A Summary of

Progress in Research and Management." Trout Unlimited, http://whirlingdisease.montana.edu/pdfs/TU_Report_99.pdf (accessed 4 December 2008).

Nico, L. G., and P. L. Fuller. "Spatial and Temporal Patterns of Nonindigenous Fish Introductions in the United States." *Fisheries* 24, no. 1 (1999): 16–27.

Nielsen, J. L., K. D. Crow, and M. C. Fountain. "Microsatellite Diversity and Conservation of a Relic Trout Population: McCloud River Redband Trout." *Molecular Ecology* 8 (1999): s129–s142.

Nielsen, L. A. "History of Inland Fisheries Management in North America." In *Inland Fisheries Management in North America*, edited by C. C. Kohler and W. A. Hubert, 3–31. Bethesda: American Fisheries Society, 1993.

Oakes, R. A. *Genealogical and Family History of the County of Jefferson, New York*. New York: Lewis, 1905.

O'Brien, S. J., and E. Mayr, "Bureaucratic Mischief: Recognizing Endangered Species and Subspecies." *Science* 251 (1991): 1187–1188.

O'Grodnick, J. J. "Susceptibility of Various Salmonids to Whirling Disease (*Myxosoma cerebralis*)." *Transactions of the American Fisheries Society* 108 (1979): 187–190.

Osborne, M. A. "Acclimatizing the World: A History of the Paradigmatic Colonial Science." *Osiris* 15 (2001): 135–151.

Page, L. M., and B. M. Burr. *Freshwater Fishes*. New York: Houghton Mifflin, 1991.

Pascual, M., and M. Kinnison. "First Documented Case of Anadromy in a Population of Introduced Rainbow Trout in Patagonia, Argentina: Response to Comment." *Transactions of the American Fisheries Society* 131 (2002): 585–588.

Paul, A. J. "Can Anglers Influence the Abundance of Native and Nonnative Salmonids in a Stream from the Canadian Rocky Mountains?" *North American Journal of Fisheries Management* 23 (2003): 109–119.

Pendleton, L. H., and R. Mendelsohn. "Estimating the Economic Impact of Climate Change on the Freshwater Sportsfisheries of the Northeastern U.S." *Land Economics* 74, no. 4 (1998): 483–496.

Persky, J. "Retrospectives: Cost-Benefit Analysis and the Classical Creed." *The Journal of Economic Perspectives* 15, no. 4 (2001): 199–208.

Petersen, S. C. *The Modern Ark: A History of the Endangered Species Act*. Ph.D. diss. University of Wisconsin, Madison, 2000.

Pilliod, D. S., and C. R. Peterson. "Local and Landscape Effects of Introduced Trout on Amphibians in Historically Fishless Watersheds." *Ecosystems* 4, no. 4 (2001): 322–333.

Pisani, D. J. "Fish Culture and the Dawn of Concern over Water Pollution in the United States." *Environmental Review* 8 (1984): 117–131.

Pister, E. P. "The Desert Fishes Council: Catalyst for Change." In *Battle Against Extinction: Native Fish Management in the American West*, edited by W. L. Minckley and J. E. Deacon, 55–68. Tucson: University of Arizona Press, 1991.

Pister, E. P. "Wilderness Fish Stocking: History and Perspective." *Ecosystems* 4 (2001): 279–286.

Plutarch. *Antony*. Translated by John Dryden. Massachusetts Institute of Technology, http://classics.mit.edu/Plutarch/antony.html (accessed 3 December 2008).

Post, G. "A Simple Chemical Test for Rotenone in Water." *Progressive Fish Culturist* 17, no. 4 (1955): 190–191.

Post, J. R., M. Sullivan, S. Cox, N. P. Lester, C. J. Walters, E. A. Parkinson, A. J. Paul, L. Jackson, and B. J. Shuter. "Canada's Recreational Fisheries: The Invisible Collapse." *Fisheries* 27, no. 1 (2002): 6–17.

Potera, C. "Fishing for Answers to Whirling Disease." *Science* 278, no. 5336 (1997): 225–226.

Prevost, G. "Experimental Stocking of Speckled Trout from the Air." *Transactions of the American Fisheries Society* 65, no. 1 (1935): 277–278.

Prevost, G., and L. Piché. "Observations on the Respiration of Trout Fingerlings and a New Method of Transporting Speckled Trout (*Salvelinus fontinalis*)." *Transactions of the American Fisheries Society* 68, no. 1 (1939): 344–353.

Propst, D. B., and D. G. Gavrilis. "Role of Economic Impact Assessment Procedures in Recreational Fisheries Management." *Transactions of the American Fisheries Society* 116, no. 3 (1987): 450–460.

Prosek, J. *The Complete Angler: A Connecticut Yankee Follows in the Footsteps of Walton*. New York: HarperCollins, 1999.

Prosek, J. *Fly-Fishing the 41st: From Connecticut to Mongolia and Home Again: A Fisherman's Odyssey*. New York: HarperCollins, 2003.

Prosek, J. *Trout: An Illustrated History*. New York: Knopf, 1996.

Prosek, J. *Trout of the World*. New York: Stewart, Tabori and Chang, 2003.

Quartarone, F. *Historical Accounts of Upper Colorado River Basin Endangered Fish*. Denver: U.S. Fish and Wildlife Service, 1995.

Radonski, G. C., and A. J. Loftus. "Fish Genetics, Fish Hatcheries, Wild Fish, and Other Fables." In *Uses and Effects of Cultured Fishes in Aquatic Ecosystems*, American Fisheries Society Symposium 15, edited by H. L. Schramm, Jr., and R. G. Piper, 1–4. Bethesda: American Fisheries Society, 1995.

Radonski, G. C., N. S. Prosser, R. G. Martin, and R. H. Stroud. "Exotic Fishes and Sport Fishing." In *Distribution, Biology, and Management of Exotic Fishes*, edited by W. R. Courtenay, Jr., and J. R. Stauffer, Jr., 313–321. Baltimore: Johns Hopkins University Press, 1984.

Rahel, F. J. "Homogenization of Fish Faunas Across the United States." *Science* 288 (2000): 854–856.

Rahel, F. J. "Homogenization of Freshwater Faunas." *Annual Review of Ecology and Systematics* 33 (2002): 291–315.

Rahel, F. J. "Unauthorized Fish Introductions: Fisheries Management of the People, for the People, or by the People." In *Propagated Fish in Resource Management*, American Fisheries Society Symposium 44, edited by M. J. Nickum, P. M. Mazik, J. G. Nickum, and D. D. MacKinlay, 431–443. Bethesda: American Fisheries Society, 2004.

Rawson, T. M. *"In Common with All Citizens": Sportsmen, Indians, Fish, and Conservation in Oregon and Washington.* Ph.D. diss. University of Oregon, 2002.

Raymond, F. E. "Livingston Stone: Pioneer Fisheries Scientist; His Career in California." *The American Fly Fisher* 16, no. 1 (1990): 18–22.

Raymond, S. *Steelhead Country.* New York: Lyons and Burford, 1991.

Regenthal, A. F. "Treatment Complete." *Utah Fish and Game* 18, no. 11 (1 November 1962): 3–5.

Regier, H. A., and V. C. Applegate. "Historical Review of the Management Approach to Exploitation and Introduction in SCOL Lakes." *Journal of the Fisheries Research Board of Canada* 29 (1972): 683–692.

Reiger, J. F. *American Sportsmen and the Origins of Conservation.* 3rd ed. Corvallis: Oregon State University Press, 2001.

Rhymer, J. M., and D. Simberloff. "Extinction by Hybridization and Introgression." *Annual Review of Ecology and Systematics* 27 (1996): 83–109.

Ricciardi, A., and J. B. Rasmussen. "Extinction Rates of North American Freshwater Fauna." *Conservation Biology* 13, no. 5 (1999): 1220–1222.

Richter, B. D., D. P. Braun, M. A. Mendelson, and L. L. Master. "Threats to Imperiled Freshwater Fauna." *Conservation Biology* 11, no. 5 (1997): 1081–1093.

Ring, R. "The West's Fisheries Spin Out of Control." *High Country News* 27, no. 17 (1995).

Rinne, J. N., and J. Janisch. "Coldwater Fish Stocking and Native Fishes in Arizona: Past, Present, and Future." In *Uses and Effects of Cultured Fishes in Aquatic Ecosystems*, American Fisheries Society Symposium 15, edited by H. L. Schramm, Jr., and R. G. Piper, 397–406. Bethesda: American Fisheries Society, 1995.

Rinne, J. N., L. Riley, R. Bettaso, R. Sorenson, and K. Young. "Managing Southwestern Native and Nonnative Fishes: Can We Mix Oil and Water and Expect a Favorable Solution?" In *Propagated Fish in Resource Management*, American Fisheries Society Symposium 44, edited by M. J. Nickum, P. M. Mazik, J. G. Nickum, and D. D. MacKinlay, 445–466. Bethesda: American Fisheries Society, 2004.

Rinne, J. N., and J. A. Stefferud. "Single Versus Multiple Species Manage-
    ment: Native Fishes in Arizona." *Forest Ecology and Management* 114
    (1999): 357–365.
Robins, C. R. Untitled sidebar. *Fisheries* 14, no. 1 (1989): 5.
Rocca, A. M. *Where the Valley Meets the Mountains*. Carlsbad, Calif.: Heri-
    tage Media, 2001.
Roosevelt, R. B. "California Mountain Trout." *New York Times*, 20 May
    1879.
Roosevelt, R. B. *Fish Culture Compared in Importance with Agriculture:
    Speech of Hon. Robert B. Roosevelt, of New York, in the House of Represen-
    tatives, May 13, 1872*. Washington: F. and J. Rives and Geo A. Bailey,
    1872.
Roosevelt, T., and G. B. Grinnell, eds. *American Big-Game Hunting: The
    Book of the Boone and Crockett Club*. New York: Forest and Stream, 1893.
Rosenlund, B. D., C. Kennedy, and K. Czarnowski, "Fisheries and
    Aquatic Management, Rocky Mountain National Park, 2001." U.S.
    Department of Interior, 2001.
Rubidge, E., P. Corbett, and E. B. Taylor. "A Molecular Analysis of Hy-
    bridization Between Native Westslope Cutthroat Trout and Intro-
    duced Rainbow Trout in Southeastern British Columbia, Canada."
    Journal of Fish Biology 59, suppl. a (2001): 42–54.
Rubidge, E. M., and E. B. Taylor. "Hybrid Zone Structure and the Poten-
    tial Role of Selection in Hybridizing Populations of Native Westslope
    Cutthroat Trout (*Oncorhynchus clarki lewisi*) and Introduced Rainbow
    Trout (*O. mykiss*)." *Molecular Ecology* 13 (2004): 3735–3749.
Ryce, E. K., A. V. Zale, and R. B. Nehring. "Lack of Selection for Resis-
    tance to Whirling Disease Among Progeny of Colorado River Rain-
    bow Trout." *Journal of Aquatic Animal Health* 12 (2001): 63–68.
Saldana, L. "Aerial Fish Plant Thrills Anglers." *Los Angeles Times*, 2 Au-
    gust 1970.
Saunders, D. L., J. J. Meeuwig, and A. C. J. Vincent. "Freshwater Pro-
    tected Areas: Strategies for Conservation." *Conservation Biology* 16,
    no. 1 (2002): 30–41.
Saunderson, H. H. *History of Charlestown, New-Hampshire*. Claremont,
    N.H.: Claremont Manufacturing, 1876.
Schade, C. B., and S. A. Bonar. "Distribution and Abundance of Non-
    native Fishes in Streams of the Western United States." *North American
    Journal of Fisheries Management* 25 (2005): 1386–1394.
Schley, B. "Somewhere a River Begins." In *Sport Fishing USA*, edited by
    D. Saults and M. Walker, 211–219. Washington: GPO, 1971.
Schramm, H. L., and R. G. Piper, eds. *Uses and Effects of Cultured Fishes in
    Aquatic Ecosystems*, American Fisheries Society Symposium 15, Be-
    thesda: American Fisheries Society, 1995.

Schullery, P. *American Fly Fishing*. New York: Lyons and Burford, 1987.

Seamans, A. L. "Mr Livingston Stone." *The Covered Wagon*, May 1948: 7–10.

Shafer, C. S. "Angling with the Bark On." *Forest and Stream*, August 1921.

Shanks, W. F. G. "Fish Culture in America." *Harper's New Monthly Magazine*, November 1868: 721–739.

Shebley, W. H. "History of the Introduction of Food and Game Fishes into the Waters of California." *California Fish and Game* 3, no. 1 (1917): 3–12.

Sheehy, C. J. "American Angling: The Rise of Urbanism and the Romance of the Rod and Reel." In *Hard at Play*, edited by Kathryn Grover. Amherst: University of Massachusetts Press, 1992.

Shepard, B. B., B. E. May, and W. Urie. "Status and Conservation of Westslope Cutthroat Trout Within the Western United States." *North American Journal of Fisheries Management* 25 (2005): 1426–1440.

Sherer, T. B., R. Betarbet, C. M. Testa, B. B. Seo, J. R. Richardson, J. H. Kim, G. W. Miller, T. Yagi, A. Matsuno-Yagi, and J. T. Greenamyre. "Mechanism of Toxicity in Rotenone Models of Parkinson's Disease." *Journal of Neuroscience* 23, no. 34 (2003): 10756–10764.

Shetter, D. S. "Further Results from Spring and Fall Plantings of Legal-Sized, Hatchery-Reared Trout in Streams and Lakes of Michigan." *Transactions of the American Fisheries Society* 74 (1947): 35–58.

Shoemaker, C. D., I. N. Gabrielson, and W. L. McAtee. "Frederic Collin Walcott, 1869–1949." *Journal of Wildlife Management* 14, no. 1 (1950): 100–102.

Sierra Nevada Aquatic Research Lab. "Sierra Nevada Aquatic Research Lab." University of California, http://vesr.ucnrs.org/pages/snarlhistory.html (accessed 8 December 2008).

Signor, J. R. *Southern Pacific's Shasta Division*. Wilton, Calif.: Signature, 2000.

Sjovold, C. "'An Angling People': Nature, Sport, and Conservation in Nineteenth-Century America." Ph.D. diss. University of California, Davis, 1999.

Smiley, C. W. "Brief Notes upon Fish and Fisheries." In *Bulletin of the United States Fish Commission* 4 (1884): 359–368. Washington: GPO, 1884.

Smiley, C. W. "A Statistical Review of the Production and Distribution to Public Waters of Young Fish, by the United States Fish Commission, from Its Organization in 1871 to the Close of 1880." In *United States Commission of Fish and Fisheries, Report of the Commissioner for 1881*, 825–916. Washington: GPO, 1884.

Smith, A. W. "The Green River Scandal." *National Parks Magazine*, January 1963.

Smith, G. R., and R. F. Stearley. "The Classification and Scientific Names of Rainbow and Cutthroat Trouts." *Fisheries* 14, no. 1 (1989): 4–10.

Smith, O. R., and P. R. Needham. "Problems Arising from the Transplantation of Trout in California." *California Fish and Game* 28, no. 1 (1942): 22–27.

Smith, T. D. 1994. *Scaling Fisheries: The Science of Measuring the Effects of Fishing, 1855–1955.* Cambridge: Cambridge University Press, 1994.

Soon, C. S., M. Brass, H. Heinze, and J. Haynes. "Unconscious Determinants of Free Decisions in the Human Brain." *Nature Neuroscience* 11, no. 5 (2008): 543–545.

Sport Fishing Institute. "The Dominating Influence of Impoundment." *SFI Bulletin* 139 (1963): 1–2.

Starr, K. *California.* New York: Modern Library, 2005.

Stearley, R. F., and G. R. Smith. "Phylogeny of the Pacific Trouts and Salmons (*Oncorhynchus*) and Genera of the Family Salmonidae." *Transactions of the American Fisheries Society* 122, no. 1 (1993): 1–33.

Stebbins, R. C. *Western Reptiles and Amphibians.* 2nd ed. New York: Houghton Mifflin, 1985.

Steller, G. W. *Steller's History of Kamchatka.* Edited by M. W. Falk, translated by M. Engel and K. Willmore. Fairbanks: University of Alaska Press, 2003.

Stewart, H. "Twelve Years Experience with Rainbow Trout." *Forest and Stream,* 3 July 1897.

Stickney, R. R. *Aquaculture in the United States: A Historical Survey.* New York: John Wiley and Sons, 1996.

Stoddard, J. L., D. V. Peck, S. G. Paulsen, J. Van Sickle, C. P. Hawkins, A. T. Herlihy, R. M. Hughes, P. R. Kaufmann, D. P. Larsen, G. Lomnicky, A. R. Olsen, S. A. Peterson, P. L. Ringold, and T. R. Whittier. "An Ecological Assessment of Western Streams and Rivers." EPA 620/R-05/005, Washington: U.S. Environmental Protection Agency, 2005.

Stone, L. "The Artificial Propagation of Salmon on the Pacific Coast of the United States, with Notes on the Natural History of the Quinnat Salmon." In *Bulletin of the U.S. Fish Commission* 16 (1896): 203–235. Washington: GPO.

Stone, L. "Brook Trout and Rainbow Trout." *Forest and Stream,* 30 November 1882.

Stone, L. *Domesticated Trout.* 6th ed. Cape Vincent, N.Y.: n.p. 1901.

Stone, L. "Report of Operations During 1872 at the United States Hatching Establishment on the McCloud River, and on the California Salmonidae Generally, with a List of Specimens Collected." In *United States Commission of Fish and Fisheries, Report of the Commissioner for 1872 and 1873,* 168–215. Washington: GPO, 1874.

Stone, L. "Report of Operations at the United States Salmon-Breeding Station on the McCloud River, California, During the Season of 1879." In *United States Commission of Fish and Fisheries, Report of the Commissioner for 1879*, 695–708. Washington: GPO, 1882.

Stone, L. "Report of Operations at the United States Trout Ponds, Mc-Cloud River, California, During the Season of 1879." In *United States Commission of Fish and Fisheries, Report of the Commissioner for 1879*, 715–720. Washington: GPO, 1882.

Stone, L. "Report of Operations at the United States Trout Ponds on the M'Cloud River, Cal., During the Season of 1880." In *United States Commission of Fish and Fisheries, Report of the Commissioner for 1882*, 615–621. Washington: GPO, 1883.

Stone, L. "Report of Operations at the Trout-Breeding Station of the United States Fish Commission on the M'Cloud River, California, During the Year 1882." In *United States Commission of Fish and Fisheries, Report of the Commissioner for 1882*, 851–856. Washington: GPO, 1884.

Stone, L. "Report of Operations at the McCloud River Trout Pond Station, California, from 1 January 1st 1888 to the Closing of the Station in June 1888." NA RG 22(117).

Stone, L. "Some Brief Reminiscences of the Early Days of Fish Culture in the United States." In *Bulletin of the United States Fish Commission* 17 (1897), 337–343. Washington: GPO, 1898.

Stone, L. "Trout Culture." *Transactions of the American Fisheries Society* 2 (1872): 46–56.

Stone, M. D. "Fish Stocking Programs in Wyoming: A Balanced Perspective." In *Uses and Effects of Cultured Fishes in Aquatic Ecosystems*, American Fisheries Society Symposium 15, edited by H. L. Schramm, Jr., and R. G. Piper, 47–51. Bethesda: American Fisheries Society, 1995.

Stone, R. "Tri-State Treatment." *Utah Fish and Game* 18, no. 9 (1 September 1962): 10–11.

Suckley, G. "On the North American Species of Salmon and Trout." Reprinted in USFC *Report of the Commissioner for 1872 and 1873*, edited by S. F. Baird, 92–160, Washington: GPO, 1874.

Tattersall, I. *Becoming Human: Evolution and Human Uniqueness.* New York: Harcourt, 1999.

Taylor, J. E., III. *Making Salmon: An Environmental History of the Northwest Fisheries Crisis.* Seattle: University of Washington Press, 1999.

Taylor, J. N., W. R. Courtenay, and J. A. McCann. "Known Impacts of Exotic Fishes in the Continental United States." In *Distribution, Biology, and Management of Exotic Fishes*, edited by W. R. Courtenay, Jr., and J. R. Stauffer, Jr., 322–373. Baltimore: Johns Hopkins University Press, 1984.

Thompson, P. E. "The First Fifty Years—The Exciting Ones." In *A Century of Fisheries in North America*, edited by N. G. Benson, 1–11. Washington: American Fisheries Society, 1970.

Titcomb, J. W. "Report on the Propagation and Distribution of Food Fishes." In *U.S. Commission of Fish and Fisheries, Report of the Commissioner for the Year Ending June 30, 1903*, 29–74. Washington: GPO, 1905.

Tol, D., and J. French. "Status of a Hybridized Population of Alvord Cutthroat Trout from Virginia Creek, Nevada." In *Status and Management of Interior Stocks of Cutthroat Trout*, American Fisheries Society Symposium 4, edited by R. E. Greswell, 116–120. Bethesda: American Fisheries Society, 1988.

Towle, J. C. "Authored Ecosystems: Livingston Stone and the Transformation of California Fisheries." *Environmental History* 5, no. 1 (2000): 54–74.

Trefethen, J. B. *An American Crusade for Wildlife*. New York: Winchester, 1975.

Trout. "A Trout Unlimited Retrospective: 1959–1999." Summer 1999: 30–38.

Trout Unlimited. "Trout Unlimited Annual Report, 2006." Trout Unlimited, www.tu.org/atf/cf/%7BED0023C4-EA23–4396-9371-8509DC5B4953%7D/2006annual_f.pdf (accessed 3 December 2008).

Tunison, A. V., S. M. Mullin, and O. L. Meehan. "Extended Survey of Fish Culture in the United States." *Progressive Fish Culturist* 11, no. 4 (1949): 253–262.

Tunison, A. V., S. M. Mullin, and O. L. Meehan. "Survey of Fish Culture in the United States." *Progressive Fish Culturist* 11, no. 1 (1949): 31–69.

Turner, F. J. "The Significance of the Frontier in American History." *The Annual Report of the American Historical Association* (1893): 199–227.

Tyus, H. M., and J. F. Saunders, III. "Nonnative Fish Control and Endangered Fish Recovery: Lessons from the Colorado River." *Fisheries* 25, no. 9 (2000): 17–24.

Udall, S. L. "Foreword." In *Battle Against Extinction: Native Fish Management in the American West*, edited by W. L. Minckley and J. E. Deacon, ix–xi. Tucson: University of Arizona Press, 1991.

Udall, S. L. *The Quiet Crisis*. New York: Holt, Rinehart, and Winston, 1963.

Udall, S. L. "Review of Green River Eradication Program." 25 March 1963. NARM RG22.

U.S. Department of Interior. "Interior Department Steps Up Fight to Save Near-Extinct Wildlife." Press Release. 6 July 1964. SIA T89021, Box 7.

U.S. Department of Interior. "Janzen to Develop Program to Combat

Wildlife Extinction; Gottschalk Named Director of Sport Fisheries and Wildlife." Press Release, 2 September 1964. SIA T89021, Box 7.

U.S. Department of the Interior. *Measures to Improve Sport Fishing at Flaming Gorge Unit, Colorado River Storage Project, Wyoming and Utah.* Washington, D.C.: U.S. Department of the Interior, 13 February 1963. NARM RG22.

U.S. Fish Commission. *U.S. Commission of Fish and Fisheries, Report of the Commissioner.* Years 1871–1940. Washington: GPO. For the sake of brevity I have included all U.S. Fish Commission Reports in this citation. These reports often include an introduction by the commissioner. Instead of citing the commissioner in the text, e.g. "Baird," I have simply cited them as "USFC, *Annual Report*" and the year. However, those sections of these reports that are explicitly credited to another author, e.g. "Stone," are cited under the author's last name. As of December 2008, all of these reports are available at http://docs .lib.noaa.gov/rescue/cof/data_rescue_fish_commission_annual_ reports.html.

U.S. Fish and Wildlife Service. *2001 National Survey of Fishing, Hunting, and Wildlife-Associated Recreation.* U.S. Fish and Wildlife Service, 2001, http://permanent.access.gpo.gov/lps52835/federalaid.fws.gov/surveys/ surveys.html.

U.S. Fish and Wildlife Service. *Briefing Statement on Flaming Gorge Fishery Management Project.* 11 January 1963. NA RG 22(266).

U.S. Fish and Wildlife Service. *Colorado Squawfish Recovery Plan.* Denver: U.S. Fish and Wildlife Service, 1990.

U.S. Fish and Wildlife Service. "Committee Information Sheet." 30 January 1964. SIA T89021, Box 7.

U.S. Fish and Wildlife Service. *Economic Effects of Rainbow Trout Production by the National Fish Hatchery System: Science and Efficiency at Work for You.* U.S. Fish and Wildlife Service, 2006, www.fws.gov/southeast/ fisheries/RainbowTrout-05.pdf.

U.S. Fish and Wildlife Service. "Endangered Colorado River Basin Fish." U.S. Fish and Wildlife Service, www.fws.gov/ColoradoRiverrecovery/ Crbtc.htm (accessed 4 December 2008).

U.S. Fish and Wildlife Service. "Endangered Species Program." U.S. Fish and Wildlife Service, www.fws.gov/endangered/wildlife.html (accessed 4 December 2008).

U.S. Fish and Wildlife Service. "Endangered and Threatened Wildlife and Plants: 12-Month Finding for an Amended Petition to List the Westslope Cutthroat Trout as Threatened Throughout Its Range." *Federal Register* 65, no. 73 (14 April 2000): 20120–20123.

U.S. Fish and Wildlife Service. "Endangered and Threatened Wildlife and Plants: Reconsidered Finding for an Amended Petition to List the Westslope Cutthroat Trout as Threatened Throughout Its Range." *Federal Register* 68, no. 152 (7 August 2003): 46989–47009.

U.S. Fish and Wildlife Service. *Net Economic Values for Wildlife-Related Recreation in 2001, Addendum to the 2001 National Survey of Fishing, Hunting, and Wildlife-Associated Recreation.* U.S. Fish and Wildlife Service, 2003, http://library.fws.gov/nat_survey2001_economicvalues.pdf.

U.S. Fish and Wildlife Service. "Press Release: Interior Seeks Information on Endangered Wildlife." 27 November 1964.

U.S. Fish and Wildlife Service. *Status Review for Westslope Cutthroat Trout in the United States.* Portland: U.S. Fish and Wildlife Service, 1999.

U.S. Fish and Wildlife Service and National Oceanic and Atmospheric Administration. "Endangered and Threatened Wildlife and Plants: Proposed Policy and Proposed Rule on the Treatment of Intercrosses and Intercross Progeny (the Issue of 'Hybridization'); Request for Public Comment." *Federal Register* 61, no. 26 (7 February 1996): 4710–4713.

U.S. General Accounting Office. "National Fish Hatcheries: Authority Needed to Better Align Operations with Priorities." GAO/RCED-00-151. Washington: GPO, 2000.

U.S. General Accounting Office. "National Fish Hatcheries: Classification of the Distribution of Fish and Fish Eggs Needs Refinement." GAO/RCED-00-10, Washington: GPO, 1999.

Utah Department of Fish and Game. *Fish and Wildlife Values and Needs in Relation to National Land Reserve Resources as Influenced by Flaming Gorge Reservoir Development.* Salt Lake: Utah Department of Fish and Game Report, 1962.

Utah Department of Fish and Game. *The Impact of Flaming Gorge Unit upon Wildlife Resources in Northeastern Utah.* Salt Lake: Utah Department of Fish and Game, 1959.

Utah Department of Fish and Game and Bureau of Sport Fisheries and Wildlife. *Fish and Wildlife Values and Needs in Relation to National Land Reserve Resources as Influenced by Flaming Gorge Reservoir Development.* Utah Department of Fish and Game and Bureau of Sport Fisheries and Wildlife. 1962. NARM RG22.

Utah State Division of Wildlife Resources and Wyoming Game and Fish Commission. *Green River and Flaming Gorge Reservoir Post-Impoundment Investigations.* Utah State Division of Wildlife Resources and Wyoming Game and Fish Commission, 1971.

Utah State Government. "Utah State Fish—Bonneville Cutthroat Trout." Utah State Library, www.pioneer.utah.gov/research/utah_symbols/trout.html (accessed 2 October 2008).

Van Vooren, A. R. "The Roles of Hatcheries, Habitat, and Regulations in Wild Trout Management in Idaho." In *Uses and Effects of Cultured Fishes in Aquatic Ecosystems,* American Fisheries Society Symposium 15, edited by H. L. Schramm, Jr., and R. G. Piper, 512–517. Bethesda: American Fisheries Society, 1995.

Varley, J. D., and P. Schullery. "Yellowstone Fishes in the Mind of Man." *The American Fly Fisher* 9, no. 2 (1982): 22–28.

Vincent, E. R. "Effects of Stocking Catchable-Size Hatchery Rainbow Trout on Two Wild Trout Species in the Madison River and O'Dell Creek, Montana." *North American Journal of Fisheries Management* 7 (1987): 91–105.

Vincent, E. R. "Relative Susceptibility of Various Salmonids to Whirling Disease with Emphasis on Rainbow and Cutthroat Trout." In *Whirling Disease: Reviews and Current Topics,* American Fisheries Society Symposium 29, edited by J. L. Bartholomew and J. C. Wilson, 109–115. Bethesda: American Fisheries Society, 2002.

Vincent, D. "The Catchable Trout." *Montana Outdoors,* May–June 1972: 24–29.

Vincent, D. "The Madison River: What Does the Future Hold?" *Montana Outdoors,* July–August 1979.

Vitousek, P. M., C. M. D'Antonio, L. L. Loope, M. Rejmanek, and R. West-brooks. "Introduced Species: A Significant Component of Human-Caused Global Change." *New Zealand Journal of Ecology* 21, no. 1 (1997): 1–16.

Vredenburg, V. "Reversing Introduced Species Effects: Experimental Removal of Introduced Fish Leads to Rapid Recovery of a Declining Frog." *Proceedings of the National Academy of Sciences* 101, no. 20 (2004): 7646–7650.

Wagner, E. J. "Whirling Disease Prevention, Control, and Management: A Review." In *Whirling Disease: Reviews and Current Topics,* American Fisheries Society Symposium 29, edited by J. L. Bartholomew and J. C. Wilson, 217–225. Bethesda: American Fisheries Society, 2002.

Wagner, R. "Mission Accomplished." *Wyoming Wildlife* 26, no. 11 (1962): 12–17.

Wales, J. H. "General Report of Investigations on the McCloud River Drainage in 1938." *California Fish and Game* 25, no. 4 (1939): 272–309.

Wales, J. H. "Historical Notes on the Rainbow and Dolly Varden Trout." *The Covered Wagon,* May–June 1946.

Walker, P. G., and R. B. Nehring. "An Investigation to Determine the Cause(s) of the Disappearance of Young Wild Rainbow Trout in the Upper Colorado River, in Middle Park, Colorado." Denver: Colorado Division of Wildlife, 1995.

Walsh, M. G., and D. L. Winkelman. "Fish Assemblage Structure in an
    Oklahoma Ozark Stream Before and After Rainbow Trout Introduc-
    tion." In *Propagated Fish in Resource Management*, American Fisheries
    Society Symposium 44, edited by M. J. Nickum, P. M. Mazik, J. G.
    Nickum, and D. D. MacKinlay, 417–430. Bethesda: American Fisheries
    Society, 2004.
Walton, I., and C. Cotton. *The Complete Angler*. Renascence Editions, Uni-
    versity of Oregon, http://darkwing.uoregon.edu/~rbear/walton/index
    .html.
Waples, R. S. "Dispelling Some Myths About Hatcheries." *Fisheries* 24,
    no. 2 (1999): 12–21.
Washabaugh, W., and C. Washabaugh. *Deep Trout: Angling in Popular Cul-
    ture*. New York: Berg, 2000.
Wegener, A. *The Origin of Continents and Oceans*. Translated by J. Biram.
    New York: Dover, 1966.
Welcomme, R. L. "International Transfers of Inland Fish Species." In
    Distribution, Biology, and Management of Exotic Fishes, edited by W. R.
    Courtenay, Jr., and J. R. Stauffer, Jr., 22–40. Baltimore: Johns Hopkins
    University Press, 1984.
Wells, J. "Wild Trout Management: Commitment to Excellence." *Montana
    Outdoors*, July–August 1985: 19–22.
Whelan, G. E. "A Historical Perspective on the Philosophy Behind the
    Use of Propagated Fish in Fisheries Management: Michigan's 130-
    Year Experience." In *Propagated Fish in Resource Management*, American
    Fisheries Society Symposium 44, edited by M. J. Nickum, P. M. Mazik,
    J. G. Nickum, and D. D. MacKinlay, 307–315. Bethesda: American
    Fisheries Society, 2004.
White, R. "'Are You an Enviromentalist or Do You Work for a Living?':
    Work and Nature." In *Uncommon Ground; Rethinking the Human Place
    in Nature*, edited by W. Cronon, 171–185. New York: W. W. Norton,
    1996.
White, R. J., J. R. Karr, and W. Nehlsen. "Better Roles for Fish Stocking
    in Aquatic Resource Management." In *Uses and Effects of Cultured
    Fishes in Aquatic Ecosystems*, American Fisheries Society Symposium
    15, edited by H. L. Schramm, Jr., and R. G. Piper, 527–547. Bethesda:
    American Fisheries Society, 1995.
Whitney, A. "The Changing Role of the Fish Hatchery." *Montana Out-
    doors*, March–April 1973: 42–44.
Whitney, A. "Who Pays for What." *Montana Outdoors*, March–April 1971:
    12–15.
Wickstrom, G. M. "Bringing Back the Greenback *Oncorhynchus clarki
    stomias*." *The American Fly Fisher* 24, no. 4 (1998): 5–8.

Wiley, R. W. "A Common Sense Protocol for the Use of Hatchery-Reared Trout." In *Uses and Effects of Cultured Fishes in Aquatic Ecosystems,* American Fisheries Society Symposium 15, edited by H. L. Schramm, Jr., and R. G. Piper, 465–471. Bethesda: American Fisheries Society, 1995.

Wiley, R. W. "Diversifying Trout Fishing Opportunity in Wyoming: History, Challenges, and Guidelines." *Fisheries* 31, no. 11 (2006): 548–553.

Wiley, R. W. "Man and Native Fishes in Wyoming's Green River—A Look at History in Hindsight—The 1962 Rotenone Treatment of the Green River." Oral presentation at the Colorado Wyoming American Fisheries Society Meeting, 7 March 2006.

Wiley, R. W. "Planting Trout in Wyoming High-Elevation Wilderness Waters." *Fisheries* 28, no. 1 (2003): 22–27.

Wiley, R. W. "Trout Stocking Rates: A Wyoming Perspective." American Fisheries Society, Fisheries Management Section, Special Report 2, 1 March 2006.

Wiley, R. W., and R. S. Wydoski. "Management of Undesirable Fish Species." In *Inland Fisheries Management in North America,* edited by C. C. Kohler and W. A. Hubert, 335–354. Bethesda: American Fisheries Society, 1993.

Williams, J. E. "Genetic Purity: Saving All the Pieces." *Trout,* Spring 2006, 45.

Williams, J. E., J. E. Johnson, D. A. Hendrickson, S. Contreras-Balderas, J. D. Williams, M. Navarro-Mendoza, D. E. McAllister, and J. E. Deacon. "Fishes of North America Endangered, Threatened, or of Special Concern: 1989." *Fisheries* 14, no. 6 (1989): 2–20.

Williams, R. N., R. F. Leary, and K. P. Currens. "Localized Genetic Effects of a Long-Term Hatchery Stocking Program on Resident Rainbow Trout in the Metolius River, Oregon." *North American Journal of Fisheries Management* 17 (1997): 1079–1093.

Williams, T. "Ann and Nancy's War: Restoration of Imperiled Fish Just Got Shut Down Where It's Needed Most." *Fly Rod and Reel,* July 2005, 18–26.

Williams, T. "Bring Back the Natives: A Model for Government." *Fly Rod and Reel,* July 1997, 18–24.

Williams, T. "Disappearing Fish: Catch Some Westslope Cutthroats While You Can." *Fly Rod and Reel,* March 1999, 13–17.

Williams, T. "Environmentalists vs. Native Trout." *Fly Rod and Reel,* April 2004, 18–26.

Williams, T. "Getting Past Hatcheries: Yet Another Way Hatcheries Inflict Havoc on Wild Fish." *Fly Rod and Reel,* September 2000, 15–18.

Williams, T. "Hatchery Narcosis: Colorado Can't Seem to Shake Its 'Rubber-Fish' Habit." *Fly Rod and Reel,* March 1998, 31–36.

Williams, T. "Role Reversal on the Colorado." *Fly Rod and Reel,* April 2003, 16–21.

Williams, T. "Trout Are Wildlife, Too." *Audubon,* December 2002, available online at www.audubonmagazine.org/incite/incite0212.html.

Willmore, K., and M. Engel. "Translators' Preface." In *Steller's History of Kamchatka,* edited by M. W. Falk, translated by M. Engel and K. Willmore, ix–xiv. Fairbanks: University of Alaska Press, 2003.

Wingate, P. J. "United States View and Regulations on Fish Introductions." *Canadian Journal of Fisheries and Aquatic Sciences* 48, suppl. 1 (1991): 167–170.

Woodham Smith, C. *The Reason Why.* New York: Time, 1962.

Wyoming Game and Fish Commission. "Suggested Editorial: Wyoming Game and Fish Commission." 22 June 1962. NARM RG 22.

Yimin, Z. "Marketing of Rainbow Trout in China." Food and Agriculture Organization of the United Nations, 1994, www.fao.org/docrep/field/003/AB903E/AB903E00.htm.

Yoshiyama, R. M., and F. W. Fisher. "Long Time Past: Baird Station and the McCloud Wintu." *Fisheries* 26, no. 3 (2001): 6–22.

Zhaoming, W., and Y. Yuhui. "Cold Water Fish Culture in China." In *Coldwater Fisheries in the Trans-Himalayan Countries,* edited by T. Petr and S. B. Swar. Food and Agriculture Organization of the United Nations, 2002, www.fao.org/DOCREP/005/Y3994E/y3994e0b.htm.

Index